Encyclopaedia of
Nanotechnology

Encyclopaedia of
Nanotechnology

Vol. 6 : Energy and Environment

Dr. Rita Khare

AGROTECH PRESS
JAIPUR ■ NEW DELHI

AGROTECH PRESS

Regd. Off. "Bony Residency", Gate No. 2
Opp. Tilak Public School, Vishweswariya Nagar,
Gopalpura Road, Jaipur - 302018, India
Ph: 0141 - 2761280, Telefax: 2761381
E-mail: agrotechpress@gmail.com

Delhi Office: 102, 1st Floor, Satyam House,
4327/3, Ansari Road, Darya Ganj,
New Delhi - 110002, India
Ph: 011-45652440
visit us: www.agrotechpress.com

Distributed by :
KEYA INTERNATIONAL PRESS LLP
E-mail: keyainternationalpressllp@gmail.com

Published in India, 2024

ISBN : 9789384568566 (10 Vol. Set)

Typeset by Agrotech Press

Printed and bound in India by Replika Press Pvt. Ltd.

Contents

Preface

While nanomaterials have been a part of our everyday life for quite some time, the past two decades have witnessed a fast growth of the nanotechnology sector. Possible areas of application for nanomaterials are almost unlimited, from aircrafts to ski poles. However it is not only the strength that will be increased, but also functional properties like electronic conductivity, magnetic behaviour and optical properties. These are influenced by controlling the structure of the materials at the nanoscale level. Nanotechnology will bring us new materials with these unique properties, which make possible the development and production of new devices and products. Nanotechnology is being used in several applications to improve the environment and to produce more efficient and cost-effective energy, as generating less pollution during the manufacture of materials, producing solar cells that generate electricity at a competitive cost, cleaning up organic chemicals polluting groundwater, clearing volatile organic compounds (VOCs) from air, and so forth. The nanostructured materials can be made by using different synthesis technologies. Research has seen the beginning of what is possible to create of nanomaterials, and new techniques are constantly being developed. Development of new energy technologies and technologies for a cleaner environment are two important focus areas. For example, the development of light and strong new materials would make planes, trains and cars lighter and thus reduce energy consumption. Development of effective methods for the conversion of one type of energy to another is another important area. Materials with new functional properties will be able to streamline the energy conversion, for example from sunlight to electricity in solar cells, or from electrical energy to chemical energy in the form of hydrogen gas. The production of new and effective nanomaterials will also provide an environmental benefit because the material need will be less than with the use of traditional materials.

Nanotechnological products, processes and applications are expected to contribute significantly to environmental and climate protection by saving raw materials, energy and water as well as by reducing greenhouse gases and hazardous wastes. Using nanomaterials therefore promises certain environmental benefits and sustainability effects. Note, however, that nanotechnology currently plays a rather subordinate role in environmental protection, whether it be in research or in practical applications. Environmental engineering companies themselves attach only limited importance to nanotechnology in their respective fields.

Nanotechnology will help in identifying key research issues surrounding the implementation of environmental nanotechnologies and anticipating possible environmental impacts associated with nanomaterials. Nanomaterials exhibit special physical and chemical properties that make them interesting for novel, environmentally friendly products. It will contribute to the development of new, effective methods for purifying gases, liquids and drinking water as well as separating CO2 for storage. Nanotechnologies hold great promise for creating new means of detecting pollutants, cleaning polluted waste streams, recovering materials before they become wastes, and expanding available resources. Like all emerging technologies with great promise, the nanotechnology and nanochemistry industries will present new challenges in ensuring that environmental risks are properly managed. In the chemical industry sector, nanomaterials are applied based on their special catalytic properties in order to boost energy and resource efficiency, and nanomaterials can replace environmentally problematic chemicals in certain fields of application. High hopes are being placed in nano-technologically optimized products and processes for energy production and storage; these are currently in the development phase and are slated to contribute significantly to climate protection and solving our energy problems in the future.

The volume places emphasis on the sustainable potential of nanotechnology. Determining the actual effects of a product on the environment – both positive and negative – requires examining the entire life cycle from production of the raw material to disposal at the end of the life cycle. This book presents the possible applications of nano-sized materials in all environmental processes, providing the most reliable guideline for the selection of nanomaterials to improve the efficiency of environmental processes. It focuses on designing specific nanomaterials for environmental processes and pollutants. It presents the impact and influence of nanomaterials on the environment and discusses how to avoid causing contamination by the use of nanomaterials.

Dr. Rita Khare

1

Introduction

The environmental impact of nanotechnology is the possible effects that the use of nanotechnological materials and devices will have on the environment. As nanotechnology is an emerging field, there is great debate regarding to what extent industrial and commercial use of nanomaterials will affect organisms and ecosystems. Nanotechnology's environmental impact can be split into two aspects: the potential for nanotechnological innovations to help improve the environment, and the possibly novel type of pollution that nanotechnological materials might cause if released into the environment.

Green nanotechnology refers to the use of nanotechnology to enhance the environmental sustainability of processes producing negative externalities. It also refers to the use of the products of nanotechnology to enhance sustainability. It includes making green nano-products and using nano-products in support of sustainability. Green nanotechnology has been described as the development of clean technologies, "to minimize potential environmental and human health risks associated with the manufacture and use of nanotechnology products, and to encourage replacement of existing products with new nano-products that are more environmentally friendly throughout their lifecycle."

Green nanotechnology has two goals: producing nanomaterials and products without harming the environment or human health, and producing nano-products that provide solutions to environmental problems. It uses existing principles of green chemistry and green engineering to make nanomaterials and nano-products without toxic ingredients, at low temperatures using less energy and renewable inputs wherever possible, and using lifecycle thinking in all design and engineering stages.

In addition to making nanomaterials and products with less impact to the environment, green nanotechnology also means using nanotechnology to make current manufacturing processes for non-nano materials and products more environmentally friendly. For example, nanoscale membranes can help separate desired chemical reaction products from waste materials. Nanoscale catalysts can make chemical reactions more efficient and less wasteful. Sensors at the nanoscale can form a part of process control systems, working with nano-enabled information systems. Using alternative energy systems, made possible by nanotechnology, is another way to "green" manufacturing processes.

The second goal of green nanotechnology involves developing products that benefit the environment either directly or indirectly. Nanomaterials or products directly can clean hazardous waste sites, desalinate water, treat pollutants, or sense and monitor environmental pollutants. Indirectly, lightweight nanocomposites for automobiles and other means of transportation could save fuel and reduce materials used for production; nanotechnology-enabled fuel cells and light-emitting diodes (LEDs) could reduce pollution from energy generation and help conserve fossil fuels; self-cleaning nanoscale surface coatings could reduce or eliminate many cleaning chemicals used in regular maintenance routines; and enhanced battery life could lead to less material use and less waste. Green Nanotechnology takes a broad systems view of nanomaterials and products, ensuring that unforeseen consequences are minimized and that impacts are anticipated throughout the full life cycle.

Solar Cells

One major project that is being worked on is the development of nanotechnology in solar cells. Solar cells are more efficient as they get tinier and solar energy is a renewable resource. The price per watt of solar energy is lower than one dollar.

Nanotechnology is already used to provide improved performance coatings for photovoltaic (PV) and solar thermal panels. Hydrophobic and self-cleaning properties combine to create more efficient solar panels, especially during inclement weather. PV covered with nanotechnology coatings are said to stay cleaner for longer to ensure maximum energy efficiency is maintained.

Water Treatment

Nanotechnology offers the potential of novel nanomaterials for the treatment of surface water, groundwater and wastewater contaminated by toxic metal ions, organic and inorganic solutes and microorganisms. Due to their unique activity toward recalcitrant contaminants many nanomaterials are under active research and development for use in the treatment of water.

Pollution

Scientists have been researching the capabilities of buckminsterfullerene in controlling pollution, as it may be able to control certain chemical reactions. Buckminsterfullerene has been demonstrated as having the ability of inducing the protection of reactive oxygen species and causing lipid peroxidation. This material may allow for hydrogen fuel to be more accessible to consumers.

Thermal Electrical Energy Conversion from the Nanotechnology Perspective

The alkali metal thermal to electric converter (AMTEC), originally called the sodium heat engine (SHE) was invented by Joseph T. Kummer and Neill Weber at Ford in 1966, and is described in US Patents 3,404,036; 3,458,356; 3,535,163; and 4,049,877; it is a thermally regenerative electrochemical device for the direct conversion of heat to electrical energy. It is characterized by high potential efficiencies and no moving parts except the working fluid, which make it a candidate for space power applications.

This device accepts a heat input in a range from about 900 K–1300 K and produces direct current with predicted device efficiencies of 15-40%. In the AMTEC sodium is driven around a closed thermodynamic cycle between a high temperature heat reservoir and a cooler reservoir at the heat rejection temperature. The unique feature of the AMTEC cycle is that sodium ion conduction between a high pressure or activity region and a low pressure or activity region on either side of a highly ionically conducting refractory solid electrolyte, is thermodynamically nearly equivalent to an isothermal expansion of sodium vapor between the same high and low pressure. Electrochemical oxidation of neutral sodium at the anode leads to sodium ions which traverse the solid electrolyte and electrons which travel from the anode through an external circuit where they perform electrical work, to the low pressure cathode, where they recombine with the ions to produce low pressure sodium gas. The sodium gas generated at the cathode then travels to a condenser at the heat rejection temperature of perhaps 400–700 K where liquid sodium reforms. The AMTEC thus is an electrochemical concentration cell which converts the work generated by expansion of sodium vapor directly into electric power.

The converter is based on the electrolyte used in the sodium-sulfur battery, sodium beta"-alumina, a crystalline phase of somewhat variable composition containing aluminum oxide, Al2O3, and sodium oxide, Na2O, in a nominal ratio of 5:1, and a small amount of the oxide of a small cation metal, usually lithium or magnesium, which stabilizes the beta" crystal structure. The sodium beta"-alumina solid electrolyte (BASE) [ceramic] is nearly insulating with respect to transport of electrons, and is a thermodynamically stable phase in contact with both liquid sodium and sodium at low pressure.

Single cell AMTECs with open voltages as high as > 1.55 V and maximum power density as high as > 0.50 W/cm^2 at temperature of 1173K (900°C) have been obtained with long term stable refractory metal electrodes.

Efficiency of AMTEC cells has reached 16% in the laboratory. High voltage multi-tube modules are predicted to be 20% to 25% efficient, and power densities up to 0.2 kilowatts per liter appear to be achievable in the near future. Calculations show that replacing sodium with a potassium working fluid increases the peak efficiency from 28% to 31% at 1100 K with a 1 mm thick BASE tube.

Most work on AMTECs has concerned sodium working fluid devices. Potassium AMTECs have been run with potassium beta" alumina solid electrolyte ceramics, and show improved power at lower operating temperatures compared with sodium AMTECs.

A detailed quantitative model of the mass transport and intefacial kinetics behaviour of AMTEC electrodes has been developed and used to fit and analyze the performance of a wide variety of electrodes, and make predictions of the performance of optimized electrodes. The interfacial electrochemical kinetics can be further described quantitatively with a tunneling, diffusion, and desorption model. A reversible thermodynamic cycle for AMTEC shows that it is, at best, slightly less efficient than a Carnot cycle.

AMTEC requires energy input at modest elevated temperatures, and not at a specific wavelength, it is easily adapted to any heat source, including radioisotope, concentrated solar, external combustion, or nuclear reactor. A solar thermal power conversion system based on an AMTEC has advantages over other technologies (including photovoltaic systems) in terms of the total power that can be achieved with such a system and the simplicity of the system (which includes the collector, energy storage (thermal storage with phase change material) and power conversion in a compact unit). The overall system could achieve as high as 14 W/kg with present collector technology and future AMTEC conversion efficiencies. The energy storage system outperforms batteries, and the temperatures at which the system operates allows long life and reduced radiator size (heat reject temperature of 600 K). Deep-space applications would use radioisotope thermoelectric generators; hybrid systems are in design.

While space power systems are of intrinsic interest, terrestrial applications will offer large scale applications for AMTEC systems. At the +25% efficiency projected for the device and projected costs of $350/ kW, AMTEC is expected to prove useful for a very wide variety of distributed generation applications including self-powered fans for high efficiency furnaces and water heaters and recreational vehicle power supplies. Cathodic protection of pipelines, remote telemetry from oil well sites are other areas where this type of electrical generation might be used. The potential to scavenge waste heat may allow for integration of this technology into general residential and commercial cogeneration schemes although costs per kilowatt-hour would have to drop substantially from current projections.

Piezoelectric and Piezotronic Effects in Energy

Piezoelectricity is the electric charge that accumulates in certain solid materials (such as crystals, certain ceramics, and biological matter such as bone, DNA and various proteins) in response to applied mechanical stress. The word *piezoelectricity* means electricity resulting from pressure. It is derived from the Greek *piezo* or *piezein*, which means to squeeze or press, and *electric* or *electron*, which stands for amber, an ancient source of electric charge. Piezoelectricity was discovered in 1880 by French physicists Jacques and Pierre Curie.

The piezoelectric effect is understood as the linear electromechanical interaction between the mechanical and the electrical state in crystalline materials with no inversion symmetry. The piezoelectric effect is a reversible process in that materials exhibiting the direct piezoelectric effect (the internal generation of electrical charge resulting from an applied mechanical force) also exhibit the reverse piezoelectric effect (the internal generation of a mechanical strain resulting from an applied electrical field). For example, lead zirconate titanate crystals will generate measurable piezoelectricity when their static structure is deformed by about 0.1% of the original dimension. Conversely, those same crystals will change about 0.1% of their static dimension when an external electric field is applied to the material. The inverse piezoelectric effect is used in production of ultrasonic sound waves.

Piezoelectricity is found in useful applications such as the production and detection of sound, generation of high voltages, electronic frequency generation, microbalances, to drive an ultrasonic nozzle and ultrafine focusing of optical assemblies. It is also the basis of a number of scientific instrumental techniques with atomic resolution, the scanning probe microscopies such as STM, AFM, MTA, SNOM, etc., and everyday uses such as acting as the ignition source for cigarette lighters and push-start propane barbecues.

History

Discovery and Early Research: The pyroelectric effect, by which a material generates an electric potential in response to a temperature change, was studied by Carl Linnaeus and Franz Aepinus in the mid-18th century. Drawing on this knowledge, both René Just Haüy and Antoine César Becquerel posited a relationship between mechanical stress and electric charge; however, experiments by both proved inconclusive.

The first demonstration of the direct piezoelectric effect was in 1880 by the brothers Pierre Curie and Jacques Curie. They combined their knowledge of pyroelectricity with their understanding of the underlying crystal structures that gave rise to pyroelectricity to predict crystal behaviour, and demonstrated the effect using crystals of tourmaline, quartz, topaz, cane sugar, and Rochelle salt (sodium potassium tartrate tetrahydrate). Quartz and Rochelle salt exhibited the most piezoelectricity.

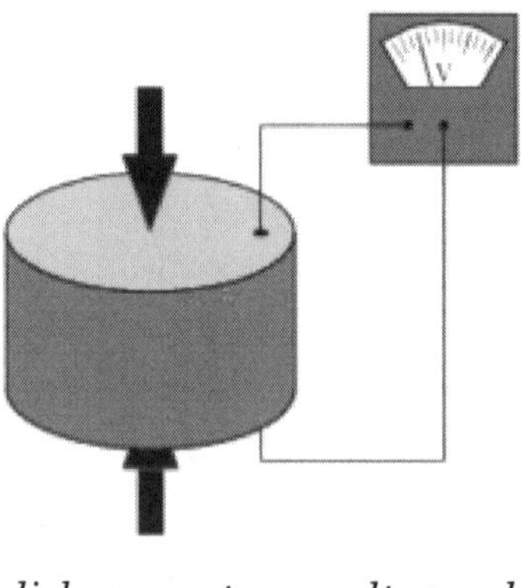

Figure: *A piezoelectric disk generates a voltage when deformed (change in shape is greatly exaggerated)*

The Curies, however, did not predict the converse piezoelectric effect. The converse effect was mathematically deduced from fundamental thermodynamic principles by Gabriel Lippmann in 1881. The Curies immediately confirmed the existence of the converse effect, and went on to obtain quantitative proof of the complete reversibility of electro-elasto-mechanical deformations in piezoelectric crystals.

For the next few decades, piezoelectricity remained something of a laboratory curiosity. More work was done to explore and define the crystal structures that exhibited piezoelectricity. This culminated in 1910 with the publication of Woldemar Voigt's *Lehrbuch der Kristallphysik*, which described the 20 natural crystal classes capable of piezoelectricity, and rigorously defined the piezoelectric constants using tensor analysis.

World War I and Post-War

The first practical application for piezoelectric devices was sonar, first developed during World War I. In France in 1917, Paul Langevin and his coworkers developed an ultrasonic submarine detector. The detector consisted of a transducer, made of thin quartz crystals carefully glued between two steel plates, and a hydrophone to detect the returned echo. By emitting a high-frequency chirp from the transducer, and measuring the amount of time it takes to hear an echo from the sound waves bouncing off an object, one can calculate the distance to that object.

The use of piezoelectricity in sonar, and the success of that project, created intense development interest in piezoelectric devices. Over the next few decades, new piezoelectric materials and new applications for those materials were explored and developed.

Piezoelectric devices found homes in many fields. Ceramic phonograph cartridges simplified player design, were cheap and accurate, and made record players cheaper to maintain and easier to build. The development of the ultrasonic transducer allowed for easy measurement of viscosity and elasticity in fluids and solids, resulting in huge advances in materials research. Ultrasonic time-domain reflectometres (which send an ultrasonic pulse through a material and measure reflections from discontinuities) could find flaws inside cast metal and stone objects, improving structural safety.

World War II and Post-War

During World War II, independent research groups in the United States, Russia, and Japan discovered a new class of synthetic materials, called ferroelectrics, which exhibited piezoelectric constants many times higher than natural materials. This led to intense research to develop barium titanate and later lead zirconate titanate materials with specific properties for particular applications.

One significant example of the use of piezoelectric crystals was developed by Bell Telephone Laboratories. Following World War I, Frederick R. Lack, working in radio telephony in the engineering department, developed the "AT cut" crystal, a crystal that operated through a wide range of temperatures. Lack's crystal didn't need the heavy accessories previous crystal used, facilitating its use on aircraft. This development allowed Allied air forces to engage in coordinated mass attacks through the use of aviation radio.

Development of piezoelectric devices and materials in the United States was kept within the companies doing the development, mostly due to the wartime beginnings of the field, and in the interests of securing profitable patents. New materials were the first to be developed — quartz crystals were the first commercially exploited piezoelectric material, but scientists searched for higher-performance materials. Despite the advances in materials and the maturation of manufacturing processes, the United States market had not grown as quickly. Without many new applications, the growth of the United States' piezoelectric industry suffered.

In contrast, Japanese manufacturers shared their information, quickly overcoming technical and manufacturing challenges and creating new markets. Japanese efforts in materials research created piezoceramic materials competitive to the U.S. materials, but free of expensive patent restrictions. Major Japanese piezoelectric developments include new designs of piezoceramic filters for radios and televisions, piezo buzzers and audio transducers that can connect directly to electronic circuits, and the piezoelectric igniter, which generates sparks for small engine ignition systems (and gas-grill lighters) by compressing a ceramic disc. Ultrasonic transducers that transmit sound waves through air had existed for quite some time, but first saw major commercial use in early television remote controls. These transducers now are mounted on several car models as an echolocation device, helping the driver determine the distance from the rear of the car to any objects that may be in its path.

Mechanism

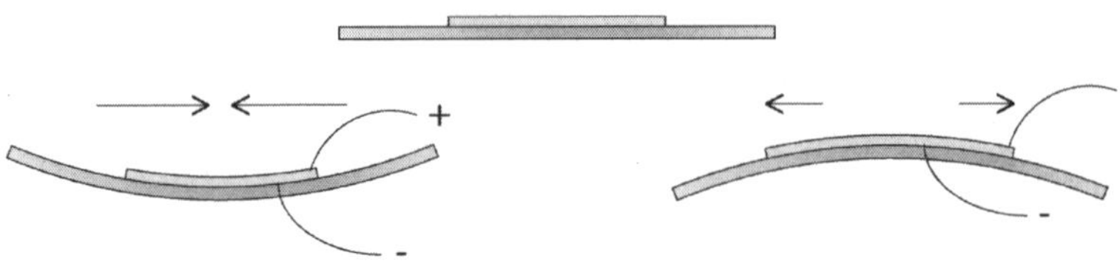

Figure: *Piezoelectric plate used to convert audio signal to sound waves*

The nature of the piezoelectric effect is closely related to the occurrence of electric dipole moments in solids. The latter may either be induced for ions on crystal lattice sites with asymmetric charge surroundings (as in $BaTiO_3$ and PZTs) or may directly be carried by molecular groups (as in cane sugar). The dipole density or polarization (dimensionality [Cm/m^3]) may easily be calculated for crystals by summing up the dipole moments per volume of the crystallographic unit cell. As every dipole is a vector, the dipole density P is a vector field. Dipoles near each other tend to be aligned in regions called Weiss domains. The domains are usually randomly oriented, but can be aligned using the process of *poling* (not the same as magnetic poling), a process by which a strong electric field is applied across the material, usually at elevated temperatures. Not all piezoelectric materials can be poled.

Of decisive importance for the piezoelectric effect is the change of polarization P when applying a mechanical stress. This might either be caused by a re-configuration of the dipole-inducing surrounding or by re-orientation of molecular dipole moments under the influence of the external stress. Piezoelectricity may then manifest in a variation of the polarization strength, its direction or both, with the details depending on 1. the orientation of P within the crystal, 2. crystal symmetry and 3. the applied mechanical stress. The change in P appears as a variation of surface charge density upon the crystal faces, i.e. as a variation of the electrical field extending between the faces caused by a change in dipole density in the bulk. For example, a 1 cm^3 cube of quartz with 2 kN (500 lbf) of correctly applied force can produce a voltage of 12500 V.

Piezoelectric materials also show the opposite effect, called converse piezoelectric effect, where the application of an electrical field creates mechanical deformation in the crystal.

Mathematical Description

Piezoelectricity is the combined effect of the electrical behaviour of the material:

$$\mathrm{D} = \varepsilon \mathrm{E} \quad \Rightarrow \quad D_i = \varepsilon_{ij} E_j$$

where D is the electric charge density displacement (electric displacement), ε is permittivity and E is electric field strength, and Hooke's Law:

$$\mathrm{S} = \mathrm{s}\mathrm{T} \quad \Rightarrow \quad S_{ij} = s_{ijkl} T_{kl}$$

where S is strain, s is compliance and T is stress.

These may be combined into so-called *coupled equations*, of which the strain-charge form is:

$$\begin{aligned} \mathrm{S} &= \mathrm{sT} + \mathrm{d}^t\mathrm{E} \;\Rightarrow\; S_{ij} = s_{ijkl}T_{kl} + d_{kij}E_k \\ \mathrm{D} &= \mathrm{dT} + \varepsilon\mathrm{E} \;\Rightarrow\; D_i = d_{ijk}T_{jk} + \varepsilon_{ij}E_j. \end{aligned}$$

In matrix form,

$$\begin{aligned} \{S\} &= \left[s^E\right]\{T\} + [d^t]\{E\} \\ \{D\} &= [d]\{T\} + \left[\varepsilon^T\right]\{E\}, \end{aligned}$$

where $[d]$ is the matrix for the direct piezoelectric effect and $[d^t]$ is the matrix for the converse piezoelectric effect. The superscript E indicates a zero, or constant, electric field; the superscript T indicates a zero, or constant, stress field; and the superscript t stands for transposition of a matrix.

The strain-charge for a material of the 4mm (C_{4v}) crystal class (such as a poled piezoelectric ceramic such as tetragonal PZT or $BaTiO_3$) as well as the 6mm crystal class may also be written as (ANSI IEEE 176):

$$\begin{bmatrix} S_1 \\ S_2 \\ S_3 \\ S_4 \\ S_5 \\ S_6 \end{bmatrix} = \begin{bmatrix} s_{11}^E & s_{12}^E & s_{13}^E & 0 & 0 & 0 \\ s_{21}^E & s_{22}^E & s_{23}^E & 0 & 0 & 0 \\ s_{31}^E & s_{32}^E & s_{33}^E & 0 & 0 & 0 \\ 0 & 0 & 0 & s_{44}^E & 0 & 0 \\ 0 & 0 & 0 & 0 & s_{55}^E & 0 \\ 0 & 0 & 0 & 0 & 0 & s_{66}^E = 2\left(s_{11}^E - s_{12}^E\right) \end{bmatrix} \begin{bmatrix} T_1 \\ T_2 \\ T_3 \\ T_4 \\ T_5 \\ T_6 \end{bmatrix} + \begin{bmatrix} 0 & 0 & d_{31} \\ 0 & 0 & d_{32} \\ 0 & 0 & d_{33} \\ 0 & d_{24} & 0 \\ d_{15} & 0 & 0 \\ 0 & 0 & 0 \end{bmatrix} \begin{bmatrix} E_1 \\ E_2 \\ E_3 \end{bmatrix}$$

$$\begin{bmatrix} D_1 \\ D_2 \\ D_3 \end{bmatrix} = \begin{bmatrix} 0 & 0 & 0 & 0 & d_{15} & 0 \\ 0 & 0 & 0 & d_{24} & 0 & 0 \\ d_{31} & d_{32} & d_{33} & 0 & 0 & 0 \end{bmatrix} \begin{bmatrix} T_1 \\ T_2 \\ T_3 \\ T_4 \\ T_5 \\ T_6 \end{bmatrix} + \begin{bmatrix} \varepsilon_{11} & 0 & 0 \\ 0 & \varepsilon_{22} & 0 \\ 0 & 0 & \varepsilon_{33} \end{bmatrix} \begin{bmatrix} E_1 \\ E_2 \\ E_3 \end{bmatrix}$$

where the first equation represents the relationship for the converse piezoelectric effect and the latter for the direct piezoelectric effect.

Although the above equations are the most used form in literature, some comments about the notation are necessary. Generally D and E are vectors, that is, Cartesian tensor of rank-1; and permittivity ε is Cartesian tensor of rank 2. Strain and stress are, in principle, also rank-2 tensors. But conventionally, because strain and stress are all symmetric tensors, the subscript of strain and stress can be re-labelled in the following fashion: $11 \to 1$; $22 \to 2$; $33 \to 3$; $23 \to 4$; $13 \to 5$; 12

→ 6. (Different convention may be used by different authors in literature. Say, some use 12 → 4; 23 → 5; 31 → 6 instead.) That is why S and T appear to have the "vector form" of 6 components. Consequently, s appears to be a 6 by 6 matrix instead of rank-4 tensor. Such a re-labelled notation is often called Voigt notation.

In total, there are 4 piezoelectric coefficients, d_{ij}, e_{ij}, g_{ij}, and h_{ij} defined as follows:

$$d_{ij} = \left(\frac{\partial D_i}{\partial T_j}\right)^E = \left(\frac{\partial S_j}{\partial E_i}\right)^T$$

$$e_{ij} = \left(\frac{\partial D_i}{\partial S_j}\right)^E = -\left(\frac{\partial T_j}{\partial E_i}\right)^S$$

$$g_{ij} = -\left(\frac{\partial E_i}{\partial T_j}\right)^D = \left(\frac{\partial S_j}{\partial D_i}\right)^T$$

$$h_{ij} = -\left(\frac{\partial E_i}{\partial S_j}\right)^D = -\left(\frac{\partial T_j}{\partial D_i}\right)^S$$

where the first set of 4 terms correspond to the direct piezoelectric effect and the second set of 4 terms correspond to the converse piezoelectric effect. A formalism has been worked out for those piezoelectric crystals, for which the polarization is of the crystal-field induced type, that allows for the calculation of piezoelectrical coefficients d_{ij} from electrostatic lattice constants or higher-order Madelung constants.

Crystal Classes

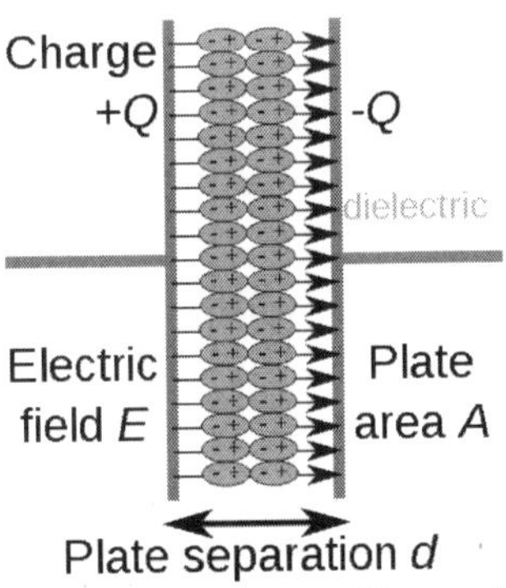

Figure: *Any spatially separated charge will result in an electric field, and therefore an electric potential. Shown here is a standard dielectric in a capacitor. In a piezoelectric device, mechanical stress, instead of an externally applied voltage, causes the charge separation in the individual atoms of the material, .*

Of the thirty-two crystal classes, twenty-one are non-centrosymmetric (not having a centre of symmetry), and of these, twenty exhibit direct piezoelectricity (the 21st is the cubic class 432). Ten of these represent the polar crystal classes, which show a spontaneous polarization without mechanical stress due to a non-vanishing electric dipole moment associated with their unit cell, and which exhibit pyroelectricity. If the dipole moment can be reversed by the application of an electric field, the material is said to be ferroelectric.

- Polar crystal classes: 1, 2, m, mm2, 4, 4 mm, 3, 3m, 6, 6 mm.
- Piezoelectric crystal classes: 1, 2, m, 222, mm2, 4, 4, 422, 4 mm, 42m, 3, 32, 3m, 6, 6, 622, 6 mm, 62m, 23, 43m.

For polar crystals, for which *P* ‘‘‘ 0 holds without applying a mechanical load, the piezoelectric effect manifests itself by changing the magnitude or the direction of *P* or both. For the non-polar, but piezoelectric crystals, on the other hand, a polarization *P* different from zero is only elicited by applying a mechanical load. For them the stress can be imagined to transform the material from a non-polar crystal class ($P = 0$) to a polar one, having $P \neq 0$.

Materials

Many materials, both natural and synthetic, exhibit piezoelectricity:

Naturally Occurring Crystals

- Quartz
- Berlinite ($AlPO_4$), a rare phosphate mineral that is structurally identical to quartz
- Sucrose (table sugar)
- Rochelle salt
- Topaz
- Tourmaline-group minerals

The action of piezoelectricity in Topaz can probably be attributed to ordering of the (F,OH) in its lattice, which is otherwise centrosymmetric: Orthorhombic Bipyramidal (mmm). Topaz has anomalous optical properties which are attributed to such ordering.

Bone

Dry bone exhibits some piezoelectric properties. Studies of Fukada *et al.* showed that these are not due to the apatite crystals, which are centrosymmetric, thus non-piezoelectric, but due to collagen. Collagen exhibits the polar uniaxial orientation of molecular dipoles in its structure and can be considered as bioelectret, a sort of dielectric material exhibiting quasipermanent space charge and dipolar charge. Potentials are thought to occur when a number of collagen molecules are stressed in the same way displacing significant numbers of the charge carriers from the inside to the surface of the specimen. Piezoelectricity of single individual collagen fibrils was measured using piezoresponse force microscopy, and it was shown that collagen fibrils behave predominantly as shear piezoelectric materials.

The piezoelectric effect is generally thought to act as a biological force sensor. This effect was exploited by research conducted at the University of Pennsylvania in the late 1970s and early 1980s, which established that sustained application of electrical potential could stimulate both resorption and growth (depending on the polarity) of bone in-vivo. Further studies in the 1990s provided the mathematical equation to confirm long bone wave propagation as to that of hexagonal (Class 6) crystals.

Other Natural Materials

Biological materials exhibiting piezoelectric properties include:

- Tendon
- Silk
- Wood due to piezoelectric texture
- Enamel
- Dentin
- DNA
- Viral proteins, including those from bacteriophage. One study has found that thin films of M13 bacteriophage can be used to construct a piezoelectric generator sufficient to operate a liquid crystal display.

Synthetic Crystals

- Gallium orthophosphate ($GaPO_4$), a quartz analogic crystal.
- Langasite ($La_3Ga_5SiO_{14}$), a quartz analogic crystal.

Synthetic Ceramics

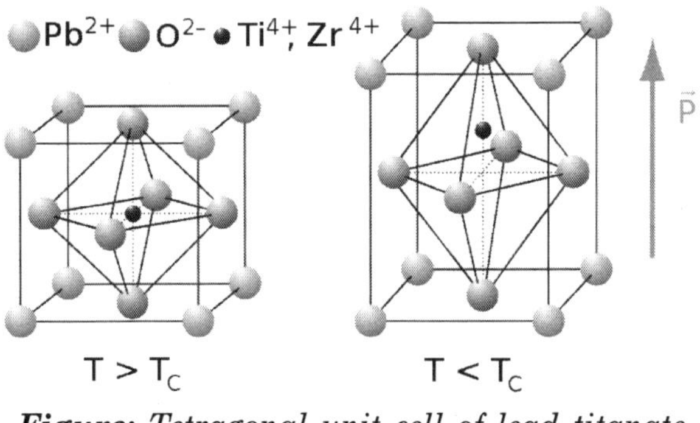

Figure: *Tetragonal unit cell of lead titanate*

The family of ceramics with perovskite or tungsten-bronze structures exhibits piezoelectricity:

- Barium titanate ($BaTiO_3$)—Barium titanate was the first piezoelectric ceramic discovered.
- Lead titanate ($PbTiO_3$)
- Lead zirconate titanate ($Pb[Zr_xTi_{1-x}]O_3$ $0 \le x \le 1$)—more commonly known as *PZT*, lead zirconate titanate is the most common piezoelectric ceramic in use today.
- Potassium niobate ($KNbO_3$)
- Lithium niobate ($LiNbO_3$)
- Lithium tantalate ($LiTaO_3$)
- Sodium tungstate (Na_2WO_3)
- Zinc oxide (ZnO)
- $Ba_2NaNb_5O_5$
- $Pb_2KNb_5O_{15}$

Lead-Free Piezoceramics

More recently, there is growing concern regarding the toxicity in lead-containing devices driven by the result of restriction of hazardous substances directive regulations. To address this concern, there has been a resurgence in the compositional development of lead-free piezoelectric materials.

- Sodium potassium niobate ($(K,Na)NbO_3$). This material is also known as NKN. In 2004, a group of Japanese researchers led by Yasuyoshi Saito discovered a sodium potassium niobate composition with properties close to those of PZT, including a high T_C. Certain compositions of this material have been shown to retain a high mechanical quality factor ($Q_m \approx 900$) with increasing vibration levels, whereas the mechanical quality factor of hard PZT degrades in such conditions. This fact makes NKN a promising replacement for high power resonance applications, such as piezoelectric transformers.
- Bismuth ferrite ($BiFeO_3$) is also a promising candidate for the replacement of lead-based ceramics.
- Sodium niobate $NaNbO_3$
- Bismuth titanate $Bi_4Ti_3O_{12}$
- Sodium bismuth titanate $Na_{0.5}Bi_{0.5}TiO_3$

So far, neither the environmental impact nor the stability of supplying these substances have been confirmed.

Polymers

- Polyvinylidene fluoride (PVDF): PVDF exhibits piezoelectricity several times greater than quartz. Unlike ceramics, where the crystal structure of the material creates the piezoelectric effect, in polymers the intertwined long-chain molecules attract and repel each other when an electric field is applied.

Organic Nanostructures

A strong shear piezoelectric activity was observed in self-assembled diphenylalanine peptide nanotubes (PNTs), indicating electric polarization directed along the tube axis. Comparison with LiNbO3 and lateral signal calibration yields sufficiently high effective piezoelectric coefficient values of at least 60 pm/V (shear response for tubes of H"200 nm in diameter). PNTs demonstrate linear deformation without irreversible degradation in a broad range of driving voltages.

Application

Currently, industrial and manufacturing is the largest application market for piezoelectric devices, followed by the automotive industry. Strong demand also comes from medical instruments as well as information and telecommunications. The global demand for piezoelectric devices was valued at approximately US$14.8 billion in 2010. The largest material group for piezoelectric devices is piezocrystal, and piezopolymer is experiencing the fastest growth due to its low weight and small size.

Piezoelectric crystals are now used in numerous ways:

High Voltage and Power Sources

Direct piezoelectricity of some substances, like quartz, can generate potential differences of thousands of volts.

- The best-known application is the electric cigarette lighter: pressing the button causes a spring-loaded hammer to hit a piezoelectric crystal, producing a sufficiently high voltage electric current that flows across a small spark gap, thus heating and igniting the gas. The portable sparkers used to ignite gas stoves work the same way, and many types of gas burners now have built-in piezo-based ignition systems.
- A similar idea is being researched by DARPA in the United States in a project called *Energy Harvesting*, which includes an attempt to power battlefield equipment by piezoelectric generators embedded in soldiers' boots. However, these energy harvesting

sources by association have an impact on the body. DARPA's effort to harness 1–2 watts from continuous shoe impact while walking were abandoned due to the impracticality and the discomfort from the additional energy expended by a person wearing the shoes. Other energy harvesting ideas include harvesting the energy from human movements in train stations or other public places and converting a dance floor to generate electricity. Vibrations from industrial machinery can also be harvested by piezoeletric materials to charge batteries for backup supplies or to power low-power microprocessors and wireless radios.

- A piezoelectric transformer is a type of AC voltage multiplier. Unlike a conventional transformer, which uses magnetic coupling between input and output, the piezoelectric transformer uses acoustic coupling. An input voltage is applied across a short length of a bar of piezoceramic material such as PZT, creating an alternating stress in the bar by the inverse piezoelectric effect and causing the whole bar to vibrate. The vibration frequency is chosen to be the resonant frequency of the block, typically in the 100 kilohertz to 1 megahertz range. A higher output voltage is then generated across another section of the bar by the piezoelectric effect. Step-up ratios of more than 1000:1 have been demonstrated. An extra feature of this transformer is that, by operating it above its resonant frequency, it can be made to appear as an inductive load, which is useful in circuits that require a controlled soft start. These devices can be used in DC-AC inverters to drive cold cathode fluorescent lamps. Piezo transformers are some of the most compact high voltage sources.

Sensors

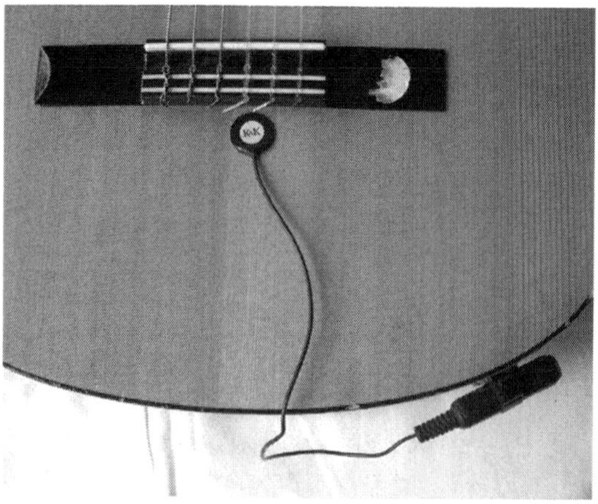

Figure: *Piezoelectric disk used as a guitar pickup*

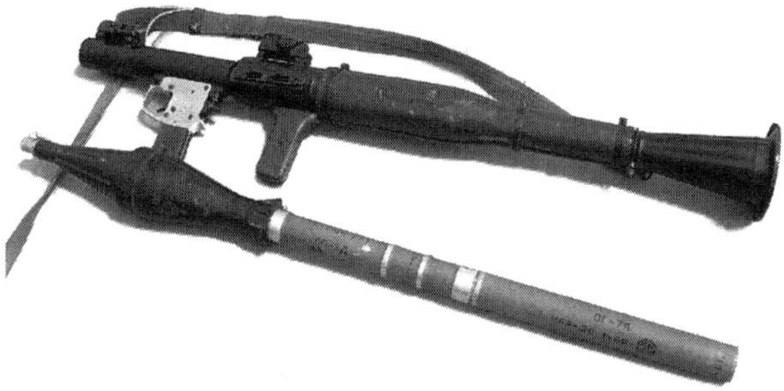

Figure: *Many rocket-propelled grenades used a piezoelectric fuse. For example: RPG-7*

The principle of operation of a piezoelectric sensor is that a physical dimension, transformed into a force, acts on two opposing faces of the sensing element. Depending on the design of a sensor, different "modes" to load the piezoelectric element can be used: longitudinal, transversal and shear. Detection of pressure variations in the form of sound is the most common sensor application, e.g. piezoelectric microphones (sound waves bend the piezoelectric material, creating a changing voltage) and piezoelectric pickups for acoustic-electric guitars. A piezo sensor attached to the body of an instrument is known as a contact microphone.

Piezoelectric sensors especially are used with high frequency sound in ultrasonic transducers for medical imaging and also industrial nondestructive testing (NDT). For many sensing techniques, the sensor can act as both a sensor and an actuator – often the term *transducer* is preferred when the device acts in this dual capacity, but most piezo devices have this property of reversibility whether it is used or not. Ultrasonic transducers, for example, can inject ultrasound waves into the body, receive the returned wave, and convert it to an electrical signal (a voltage). Most medical ultrasound transducers are piezoelectric.

In addition to those mentioned above, various sensor applications include:

- Piezoelectric elements are also used in the detection and generation of sonar waves.
- Piezoelectric materials are used in single-axis and dual-axes tilt sensing.
- Power monitoring in high power applications (e.g. medical treatment, sonochemistry and industrial processing).
- Piezoelectric microbalances are used as very sensitive chemical and biological sensors.
- Piezos are sometimes used in strain gauges.

- A piezoelectric transducer was used in the penetrometre instrument on the Huygens Probe
- Piezoelectric transducers are used in electronic drum pads to detect the impact of the drummer's sticks, and to detect muscle movements in medical acceleromyography.
- Automotive engine management systems use piezoelectric transducers to detect Engine knock (Knock Sensor, KS), also known as detonation, at certain hertz frequencies. A piezoelectric transducer is also used in fuel injection systems to measure manifold absolute pressure (MAP sensor) to determine engine load, and ultimately the fuel injectors milliseconds of on time.
- Ultrasonic piezo sensors are used in the detection of acoustic emissions in acoustic emission testing.

Actuators

Figure: *Metal disk with piezoelectric disk attached, used in a buzzer*

As very high electric fields correspond to only tiny changes in the width of the crystal, this width can be changed with better-than-μm precision, making piezo crystals the most important tool for positioning objects with extreme accuracy — thus their use in actuators. Multilayer ceramics, using layers thinner than 100 μm, allow reaching high electric fields with voltage lower than 150 V. These ceramics are used within two kinds of actuators: direct piezo actuators and Amplified piezoelectric actuators. While direct actuator's stroke is generally lower than 100 μm, amplified piezo actuators can reach millimetre strokes.

- Loudspeakers: Voltage is converted to mechanical movement of a piezoelectric polymer film.

- Piezoelectric motors: Piezoelectric elements apply a directional force to an axle, causing it to rotate. Due to the extremely small distances involved, the piezo motor is viewed as a high-precision replacement for the stepper motor.
- Piezoelectric elements can be used in laser mirror alignment, where their ability to move a large mass (the mirror mount) over microscopic distances is exploited to electronically align some laser mirrors. By precisely controlling the distance between mirrors, the laser electronics can accurately maintain optical conditions inside the laser cavity to optimize the beam output.
- A related application is the acousto-optic modulator, a device that scatters light off soundwaves in a crystal, generated by piezoelectric elements. This is useful for fine-tuning a laser's frequency.
- Atomic force microscopes and scanning tunneling microscopes employ converse piezoelectricity to keep the sensing needle close to the specimen.
- Inkjet printers: On many inkjet printers, piezoelectric crystals are used to drive the ejection of ink from the inkjet print head towards the paper.
- Diesel engines: High-performance common rail diesel engines use piezoelectric fuel injectors, first developed by Robert Bosch GmbH, instead of the more common solenoid valve devices.
- Active vibration control using amplified actuators.
- X-ray shutters.
- XY stages for micro scanning used in infrared cameras.
- Moving the patient precisely inside active CT and MRI scanners where the strong radiation or magnetism precludes electric motors.
- Crystal earpieces are sometimes used in old or low power radios.
- High-intensity focused ultrasound for localized heating or creating a localized Cavitation can be achieved, for example, in patient's body or in an industrial chemical process

Frequency Standard

The piezoelectrical properties of quartz are useful as standard of frequency.

- Quartz clocks employ a crystal oscillator made from a quartz crystal that uses a combination of both direct and converse piezoelectricity to generate a regularly timed series of electrical pulses that is used to mark time. The quartz crystal (like any elastic material) has a precisely defined natural frequency (caused by its shape and size) at which it prefers to oscillate, and this is used to stabilize the frequency of a periodic voltage applied to the crystal.

- The same principle is critical in all radio transmitters and receivers, and in computers where it creates a clock pulse. Both of these usually use a frequency multiplier to reach gigahertz ranges.

Piezoelectric Motors

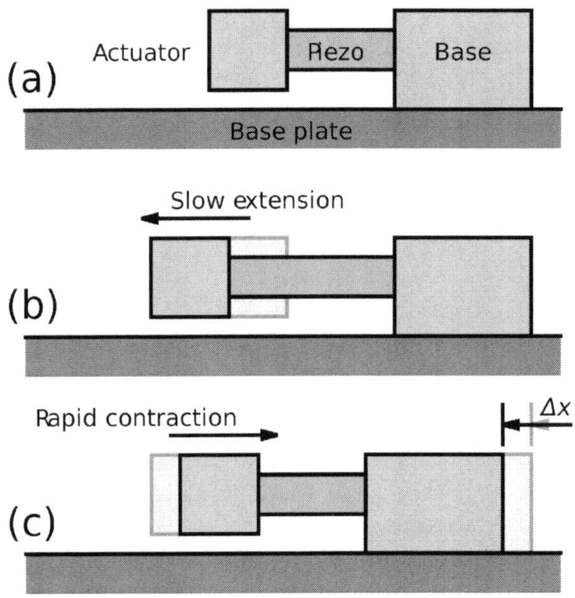

***Figure:** A slip-stick actuator.*

Types of piezoelectric motor include:

- The travelling-wave motor used for auto-focus in reflex cameras
- Inchworm motors for linear motion
- Rectangular four-quadrant motors with high power density (2.5 W/cm^3) and speed ranging from 10 nm/s to 800 mm/s.
- Stepping piezo motor, using stick-slip effect.

All these motors, except the stepping stick-slip motor work on the same principle. Driven by dual orthogonal vibration modes with a phase difference of 90°, the contact point between two surfaces vibrates in an elliptical path, producing a frictional force between the surfaces. Usually, one surface is fixed causing the other to move. In most piezoelectric motors the piezoelectric crystal is excited by a sine wave signal at the resonant frequency of the motor. Using the resonance effect, a much lower voltage can be used to produce a high vibration amplitude.

Stick-slip motor works using the inertia of a mass and the friction of a clamp. Such motors can be very small. Some are used for camera sensor displacement, thus allowing an anti-shake function.

Reduction of Vibrations and Noise

Different teams of researchers have been investigating ways to reduce vibrations in materials by attaching piezo elements to the material. When the material is bent by a vibration in one direction, the vibration-reduction system responds to the bend and sends electric power to the piezo element to bend in the other direction. Future applications of this technology are expected in cars and houses to reduce noise.

In a demonstration at the Material Vision Fair in Frankfurt in November 2005, a team from TU Darmstadt in Germany showed several panels that were hit with a rubber mallet, and the panel with the piezo element immediately stopped swinging.

Piezoelectric ceramic fibre technology is being used as an electronic damping system on some HEAD tennis rackets.

Infertility Treatment

In people with previous total fertilization failure, piezoelectric activation of oocytes together with intracytoplasmic sperm injection (ICSI) seems to improve fertilization outcomes.

Surgery

A recent application of piezoelectric ultrasound sources is piezoelectric surgery, also known as piezosurgery. Piezosurgery is a minimally invasive technique that aims to cut a target tissue with little damage to neighbouring tissues. For example, Hoigne *et al.* reported its use in hand surgery for the cutting of bone, using frequencies in the range 25–29 kHz, causing microvibrations of 60–210 μm. It has the ability to cut mineralized tissue without cutting neurovascular tissue and other soft tissue, thereby maintaining a blood-free operating area, better visibility and greater precision.

Potential Applications

Photovoltaics: The efficiency of a hybrid photovoltaic cell that contains piezoelectric materials can be increased simply by placing it near a source of ambient noise or vibration. The effect was demonstrated with organic cells using zinc oxide nanotubes. The electricity generated by the piezoelectric effect itself is a negligible percentage of the overall output. Sound levels as low as 75 decibels improved efficiency by up to 50 percent.

Efficiency peaked at 10 kHz, the resonant frequency of the nanotubes. The electrical field set up by the vibrating nanotubes interacts with electrons migrating from the organic polymer layer. This process decreases the likelihood of recombination, in which electrons are energized but settle back into a hole instead of migrating to the electron-accepting ZnO layer.

2

Nanomaterials for Fuel Cell Technologies

A fuel cell is a device that converts the chemical energy from a fuel into electricity through a chemical reaction with oxygen or another oxidizing agent. Hydrogen is the most common fuel, but hydrocarbons such as natural gas and alcohols like methanol are sometimes used. Fuel cells are different from batteries in that they require a continuous source of fuel and oxygen/air to sustain the chemical reaction whereas in a battery the chemicals present in the battery react with each other to generate an emf. Fuel cells can produce electricity continuously for as long as these inputs are supplied.

The first fuel cells were invented in 1838. The first commercial use of fuel cells came more than a century later in NASA space programs to generate power for probes, satellites and space capsules. Since then, fuel cells have been used in many other applications. Fuel cells are used for primary and backup power for commercial, industrial and residential buildings and in remote or inaccessible areas. They are also used to power fuel-cell vehicles, including forklifts, automobiles, buses, airplanes, boats, motorcycles and submarines.

There are many types of fuel cells, but they all consist of an anode, a cathode and an electrolyte that allows charges to move between the two sides of the fuel cell. Electrons are drawn from the anode to the cathode through an external circuit, producing direct current electricity. As the main difference among fuel cell types is the electrolyte, fuel cells are classified by the type of electrolyte they use followed by the difference in startup time ranging from 1 sec for PEMFC to 10 min for SOFC. Fuel cells come in a variety of sizes.

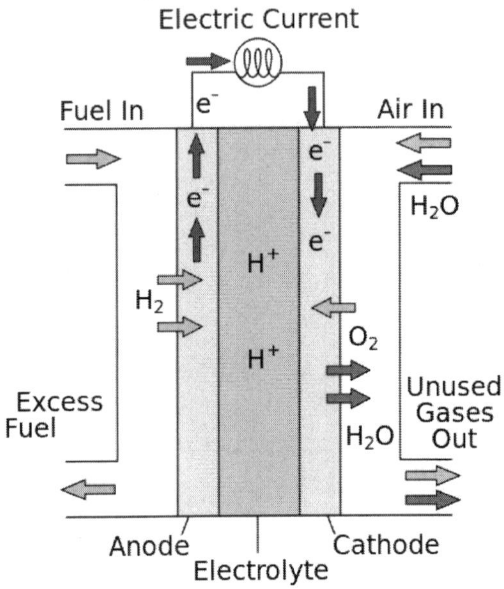

Figure: *Scheme of a proton-conducting fuel cell*

Individual fuel cells produce relatively small electrical potentials, about 0.7 volts, so cells are "stacked", or placed in series, to increase the voltage and meet an application's requirements. In addition to electricity, fuel cells produce water, heat and, depending on the fuel source, very small amounts of nitrogen dioxide and other emissions. The energy efficiency of a fuel cell is generally between 40–60%, or up to 85% efficient in cogeneration if waste heat is captured for use.

The fuel cell market is growing, and Pike Research has estimated that the stationary fuel cell market will reach 50 GW by 2020.

History

The first references to hydrogen fuel cells appeared in 1838. In a letter dated October 1838 but published in the December 1838 edition of *The London and Edinburgh Philosophical Magazine and Journal of Science*, Welsh physicist and barrister William Grove wrote about the development of his first crude fuel cells. He used a combination of sheet iron, copper and porcelain plates, and a solution of sulphate of copper and dilute acid. In a letter to the same publication written in December 1838 but published in June 1839, German physicist Christian Friedrich Schönbein discussed the first crude fuel cell that he had invented. His letter discussed current generated from hydrogen and oxygen dissolved in water. Grove later sketched his design, in 1842, in the same journal. The fuel cell he made used similar materials to today's phosphoric-acid fuel cell.

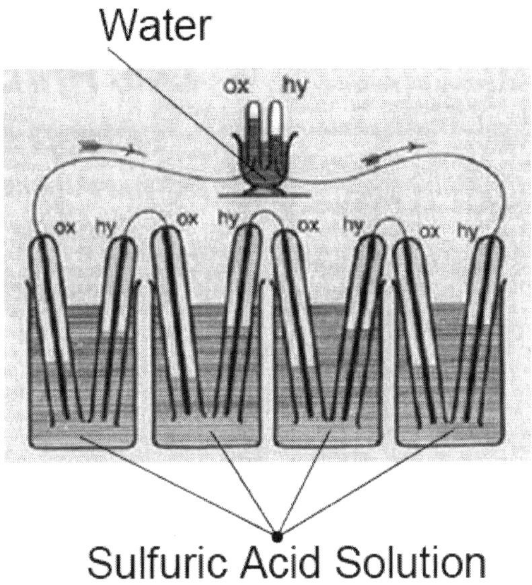

Figure: *Sketch of William Grove's 1839 fuel cell*

In 1939, British engineer Francis Thomas Bacon successfully developed a 5 kW stationary fuel cell. In 1955, W. Thomas Grubb, a chemist working for the General Electric Company (GE), further modified the original fuel cell design by using a sulphonated polystyrene ion-exchange membrane as the electrolyte. Three years later another GE chemist, Leonard Niedrach, devised a way of depositing platinum onto the membrane, which served as catalyst for the necessary hydrogen oxidation and oxygen reduction reactions. This became known as the "Grubb-Niedrach fuel cell". GE went on to develop this technology with NASA and McDonnell Aircraft, leading to its use during Project Gemini. This was the first commercial use of a fuel cell. In 1959, a team led by Harry Ihrig built a 15 kW fuel cell tractor for Allis-Chalmers, which was demonstrated across the U.S. at state fairs. This system used potassium hydroxide as the electrolyte and compressed hydrogen and oxygen as the reactants. Later in 1959, Bacon and his colleagues demonstrated a practical five-kilowatt unit capable of powering a welding machine. In the 1960s, Pratt and Whitney licensed Bacon's U.S. patents for use in the U.S. space program to supply electricity and drinking water (hydrogen and oxygen being readily available from the spacecraft tanks). In 1991, the first hydrogen fuel cell automobile was developed by Roger Billings.

UTC Power was the first company to manufacture and commercialize a large, stationary fuel cell system for use as a co-generation power plant in hospitals, universities and large office buildings. UTC Power continues to be the sole supplier of fuel cells to NASA for use in space vehicles, having supplied fuel cells for the Apollo missions, and the Space Shuttle program, and is developing fuel cells for cell phone towers and other applications.

Types of Fuel Cells; Design

Fuel cells come in many varieties; however, they all work in the same general manner. They are made up of three adjacent segments: the anode, the electrolyte, and the cathode. Two chemical reactions occur at the interfaces of the three different segments. The net result of the two reactions is that fuel is consumed, water or carbon dioxide is created, and an electric current is created, which can be used to power electrical devices, normally referred to as the load.

At the anode a catalyst oxidizes the fuel, usually hydrogen, turning the fuel into a positively charged ion and a negatively charged electron. The electrolyte is a substance specifically designed so ions can pass through it, but the electrons cannot. The freed electrons travel through a wire creating the electric current. The ions travel through the electrolyte to the cathode. Once reaching the cathode, the ions are reunited with the electrons and the two react with a third chemical, usually oxygen, to create water or carbon dioxide.

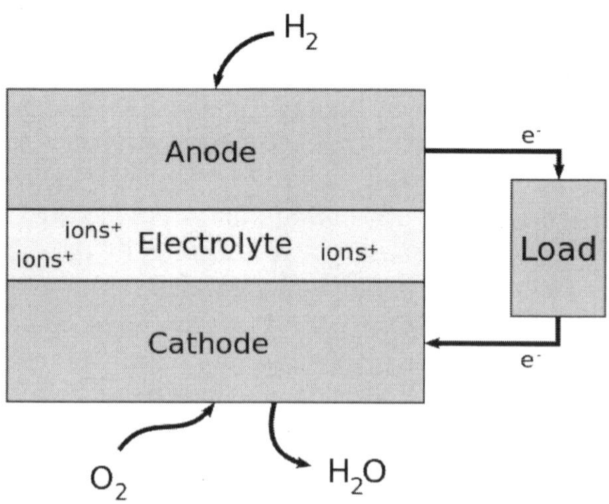

Figure: *A block diagram of a fuel cell*

The most important design features in a fuel cell are:

- The electrolyte substance. The electrolyte substance usually defines the *type* of fuel cell.
- The fuel that is used. The most common fuel is hydrogen.
- The anode catalyst breaks down the fuel into electrons and ions. The anode catalyst is usually made up of very fine platinum powder.
- The cathode catalyst turns the ions into the waste chemicals like water or carbon dioxide. The cathode catalyst is often made up of nickel but it can also be a nanomaterial-based catalyst.

A typical fuel cell produces a voltage from 0.6 V to 0.7 V at full rated load. Voltage decreases as current increases, due to several factors:

- Activation loss
- Ohmic loss (voltage drop due to resistance of the cell components and interconnections)
- Mass transport loss (depletion of reactants at catalyst sites under high loads, causing rapid loss of voltage).

To deliver the desired amount of energy, the fuel cells can be combined in series and parallel circuits to yield higher voltage, and parallel-channel of configurations allow a higher current to be supplied. Such a design is called a *fuel cell stack*. The cell surface area can be increased, to allow stronger current from each cell. In the stack, reactant gases must be distributed uniformly over all of the cells to maximize the power output.

Proton Exchange Membrane Fuel Cells (PEMFCs)

In the archetypical hydrogen–oxide proton exchange membrane fuel cell design, a proton-conducting polymer membrane (the electrolyte) separates the anode and cathode sides. This was called a "solid polymer electrolyte fuel cell" (SPEFC) in the early 1970s, before the proton exchange mechanism was well-understood. (Notice that the synonyms "polymer electrolyte membrane" and "proton exchange mechanism" result in the same acronym.)

On the anode side, hydrogen diffuses to the anode catalyst where it later dissociates into protons and electrons. These protons often react with oxidants causing them to become what are commonly referred to as multi-facilitated proton membranes. The protons are conducted through the membrane to the cathode, but the electrons are forced to travel in an external circuit (supplying power) because the membrane is electrically insulating. On the cathode catalyst, oxygen molecules react with the electrons (which have travelled through the external circuit) and protons to form water.

In addition to this pure hydrogen type, there are hydrocarbon fuels for fuel cells, including diesel, methanol and chemical hydrides. The waste products with these types of fuel are carbon dioxide and water.

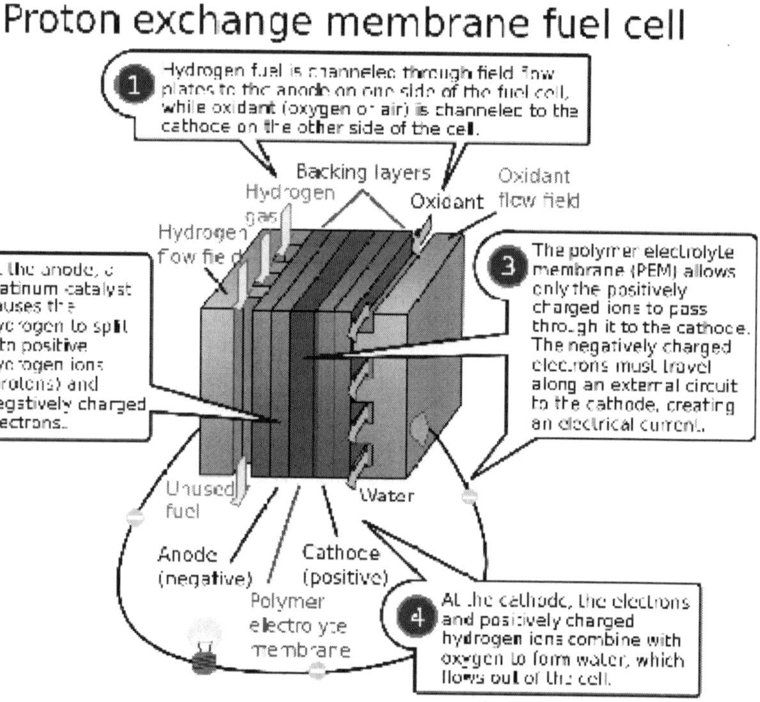

Figure: *Construction of a high-temperature PEMFC: Bipolar plate as electrode with in-milled gas channel structure, fabricated from conductive composites (enhanced with graphite, carbon black, carbon fibre, and/or carbon nanotubes for more conductivity); Porous carbon papers; reactive layer, usually on the polymer membrane applied; polymer membrane.*

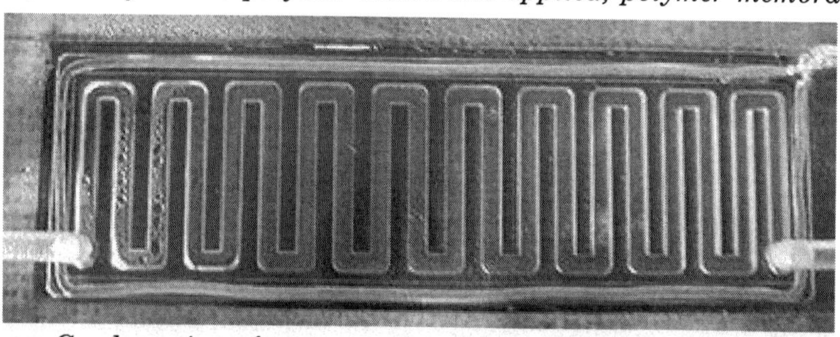

Figure: *Condensation of water produced by a PEMFC on the air channel wall. The gold wire around the cell ensures the collection of electric current.*

The different components of a PEMFC are (i) bipolar plates, (ii) electrodes, (iii) catalyst, (iv) membrane, and (v) the necessary hardware. The materials used for different parts of the fuel cells differ by type. The bipolar plates may be made of different types of materials, such as, metal, coated metal, graphite, flexible graphite, C–C composite, carbon–polymer composites etc. The membrane electrode assembly (MEA) is referred as the heart of the PEMFC and is usually made of a proton exchange membrane sandwiched between two catalyst-coated carbon papers. Platinum and/or similar type of noble metals are usually used as the catalyst for PEMFC. The electrolyte could be a polymer membrane.

Proton Exchange Membrane Fuel Cell Design Issues

- Costs. In 2013, the Department of Energy estimated that 80-kW automotive fuel cell system costs of US$67 per kilowatt could be achieved, assuming volume production of 100,000 automotive units per year and US$55 per kilowatt could be achieved, assuming volume production of 500,000 units per year. In 2008, professor Jeremy P. Meyers estimated that cost reductions over a production ramp-up period will take about 20 years after fuel-cell cars are introduced before they will be able to compete commercially with current market technologies, including gasoline internal combustion engines. Many companies are working on techniques to reduce cost in a variety of ways including reducing the amount of platinum needed in each individual cell. Ballard Power Systems has experimented with a catalyst enhanced with carbon silk, which allows a 30% reduction (1 mg/cm^2 to 0.7 mg/cm^2) in platinum usage without reduction in performance. Monash University, Melbourne uses PEDOT as a cathode. A 2011 published study documented the first metal-free electrocatalyst using relatively inexpensive doped carbon nanotubes, which are less than 1% the cost of platinum and are of equal or superior performance.
- Water and air management (in PEMFCs). In this type of fuel cell, the membrane must be hydrated, requiring water to be evaporated at precisely the same rate that it is produced. If water is evaporated too quickly, the membrane dries, resistance across it increases, and eventually it will crack, creating a gas “short circuit” where hydrogen and oxygen combine directly, generating heat that will damage the fuel cell. If the water is evaporated too slowly, the electrodes will flood, preventing the

reactants from reaching the catalyst and stopping the reaction. Methods to manage water in cells are being developed like electroosmotic pumps focusing on flow control. Just as in a combustion engine, a steady ratio between the reactant and oxygen is necessary to keep the fuel cell operating efficiently.

- Temperature management. The same temperature must be maintained throughout the cell in order to prevent destruction of the cell through thermal loading. This is particularly challenging as the $2H_2 + O_2 \rightarrow 2H_2O$ reaction is highly exothermic, so a large quantity of heat is generated within the fuel cell.
- Durability, service life, and special requirements for some type of cells. Stationary fuel cell applications typically require more than 40,000 hours of reliable operation at a temperature of –35 °C to 40 °C (–31 °F to 104 °F), while automotive fuel cells require a 5,000-hour lifespan (the equivalent of 240,000 km (150,000 mi)) under extreme temperatures. Current service life is 7,300 hours under cycling conditions. Automotive engines must also be able to start reliably at –30 °C (–22 °F) and have a high power-to-volume ratio (typically 2.5 kW per liter).
- Limited carbon monoxide tolerance of some (non-PEDOT) cathodes.

Phosphoric Acid Fuel Cell (PAFC)

Phosphoric acid fuel cells (PAFC) were first designed and introduced in 1961 by G. V. Elmore and H. A. Tanner. In these cells phosphoric acid is used as a non-conductive electrolyte to pass positive hydrogen ions from the anode to the cathode. These cells commonly work in temperatures of 150 to 200 degrees Celsius. This high temperature will cause heat and energy loss if the heat is not removed and used properly. This heat can be used to produce steam for air conditioning systems or any other thermal energy consuming system. Using this heat in cogeneration can enhance the efficiency of phosphoric acid fuel cells from 40–50% to about 80%. Phosphoric acid, the electrolyte used in PAFCs, is a non-conductive liquid acid which forces electrons to travel from anode to cathode through an external electrical circuit. Since the hydrogen ion production rate on the anode is small, platinum is used as catalyst to increase this ionization rate. A key disadvantage of these cells is the use of an acidic electrolyte. This increases the corrosion or oxidation of components exposed to phosphoric acid.

High-Temperature Fuel Cells

SOFC: Solid oxide fuel cells (SOFCs) use a solid material, most commonly a ceramic material called yttria-stabilized zirconia (YSZ), as the electrolyte. Because SOFCs are made entirely of solid materials, they are not limited to the flat plane configuration of other types of fuel cells and are often designed as rolled tubes. They require high operating temperatures (800–1000 °C) and can be run on a variety of fuels including natural gas.

SOFCs are unique in that negatively charged oxygen ions travel from the cathode (positive side of the fuel cell) to the anode (negative side of the fuel cell) instead of positively charged hydrogen ions travelling from the anode to the cathode, as is the case in all other types of fuel cells. Oxygen gas is fed through the cathode, where it absorbs electrons to create oxygen ions. The oxygen ions then travel through the electrolyte to react with hydrogen gas at the anode. The reaction at the anode produces electricity and water as by-products. Carbon dioxide may also be a by-product depending on the fuel, but the carbon emissions from an SOFC system are less than those from a fossil fuel combustion plant. The chemical reactions for the SOFC system can be expressed as follows:

Anode Reaction: $2H_2 + 2O^{2-} \rightarrow 2H_2O + 4e^-$

Cathode Reaction: $O_2 + 4e^- \rightarrow 2O^{2-}$

Overall Cell Reaction: $2H_2 + O_2 \rightarrow 2H_2O$

SOFC systems can run on fuels other than pure hydrogen gas. However, since hydrogen is necessary for the reactions listed above, the fuel selected must contain hydrogen atoms. For the fuel cell to operate, the fuel must be converted into pure hydrogen gas. SOFCs are capable of internally reforming light hydrocarbons such as methane (natural gas), propane and butane. These fuel cells are at an early stage of development.

Challenges exist in SOFC systems due to their high operating temperatures. One such challenge is the potential for carbon dust to build up on the anode, which slows down the internal reforming process. Research to address this “carbon coking” issue at the University of Pennsylvania has shown that the use of copper-based cermet (heat-resistant materials made of ceramic and metal) can reduce coking and the loss of performance. Another disadvantage of SOFC systems is slow start-up time, making SOFCs less useful for mobile applications. Despite these disadvantages, a high operating temperature provides an advantage by removing the need for a precious metal catalyst like platinum, thereby reducing cost. Additionally, waste heat from SOFC systems may be captured and reused, increasing the theoretical overall efficiency to as high as 80%–85%.

The high operating temperature is largely due to the physical properties of the YSZ electrolyte. As temperature decreases, so does the ionic conductivity of YSZ. Therefore, to obtain optimum performance of the fuel cell, a high operating temperature is required. According to their website, Ceres Power, a UK SOFC fuel cell manufacturer, has developed a method of reducing the operating temperature of their SOFC system to 500–600 degrees Celsius. They replaced the commonly used YSZ electrolyte with a CGO (cerium gadolinium oxide) electrolyte. The lower operating temperature allows them to use stainless steel instead of ceramic as the cell substrate, which reduces cost and start-up time of the system.

MCFC

Molten carbonate fuel cells (MCFCs) require a high operating temperature, 650 °C (1,200 °F), similar to SOFCs. MCFCs use lithium potassium carbonate salt as an electrolyte, and this salt liquefies at high temperatures, allowing for the movement of charge within the cell – in this case, negative carbonate ions.

Like SOFCs, MCFCs are capable of converting fossil fuel to a hydrogen-rich gas in the anode, eliminating the need to produce hydrogen externally. The reforming process creates CO_2 emissions. MCFC-compatible fuels include natural gas, biogas and gas produced from coal. The hydrogen in the gas reacts with carbonate ions from the electrolyte to produce water, carbon dioxide, electrons and small amounts of other chemicals. The electrons travel through an external circuit creating electricity and return to the cathode. There, oxygen from the air and carbon dioxide recycled from the anode react with the electrons to form carbonate ions that replenish the electrolyte, completing the circuit. The chemical reactions for an MCFC system can be expressed as follows:

Anode Reaction: $CO_3^{2-} + H_2 \rightarrow H_2O + CO_2 + 2e^-$

Cathode Reaction: $CO_2 + \frac{1}{2}O_2 + 2e^- \rightarrow CO_3^{2-}$

Overall Cell Reaction: $H_2 + \frac{1}{2}O_2 \rightarrow H_2O$

As with SOFCs, MCFC disadvantages include slow start-up times because of their high operating temperature. This makes MCFC systems not suitable for mobile applications, and this technology will most likely be used for stationary fuel cell purposes. The main challenge of MCFC technology is the cells' short life span. The high-temperature and carbonate electrolyte lead to corrosion of the anode and cathode. These factors accelerate the degradation of MCFC components, decreasing the durability and cell life. Researchers are addressing this problem by exploring corrosion-resistant materials for components as well as fuel cell designs that may increase cell life without decreasing performance.

MCFCs hold several advantages over other fuel cell technologies, including their resistance to impurities. They are not prone to "carbon coking", which refers to carbon build-up on the anode that results in reduced performance by slowing down the internal fuel reforming process. Therefore, carbon-rich fuels like gases made from coal are compatible with the system. The Department of Energy claims that coal, itself, might even be a fuel option in the future, assuming the system can be made resistant to impurities such as sulfur and particulates that result from converting coal into hydrogen. MCFCs also have relatively high efficiencies. They can reach a fuel-to-electricity efficiency of 50%, considerably higher than the 37–42% efficiency of a phosphoric acid fuel cell plant. Efficiencies can be as high as 65% when the fuel cell is paired with a turbine, and 85% if heat is captured and used in a Combined Heat and Power (CHP) system.

FuelCell Energy, a Connecticut-based fuel cell manufacturer, develops and sells MCFC fuel cells. The company says that their MCFC products range from 300 kW to 2.8 MW systems that achieve 47% electrical efficiency and can utilise CHP technology to obtain higher overall efficiencies. One product, the DFC-ERG, is combined with a gas turbine and, according to the company, it achieves an electrical efficiency of 65%.

Theoretical Maximum Efficiency

The energy efficiency of a system or device that converts energy is measured by the ratio of the amount of useful energy put out by the system ("output energy") to the total amount of energy that is put in ("input energy") or by useful output energy as a percentage of the total input energy. In the case of fuel cells, useful output energy is measured in electrical energy produced by the system. Input energy is the energy stored in the fuel. According to the U.S. Department of Energy, fuel cells are generally between 40–60% energy efficient. This is higher than some other systems for energy generation. For example, the typical internal combustion engine of a car is about 25% energy efficient. In combined heat and power (CHP) systems, the heat produced by the fuel cell is captured and put to use, increasing the efficiency of the system to up to 85–90%.

The theoretical maximum efficiency of any type of power generation system is never reached in practice, and it does not consider other steps in power generation, such as production, transportation and storage of fuel and conversion of the electricity into mechanical power. However, this calculation allows the comparison of different types of power generation. The maximum theoretical energy efficiency of a fuel cell is 83%, operating at low power density and using pure hydrogen and oxygen as reactants (assuming no heat recapture) According to the World Energy Council, this compares with a maximum theoretical efficiency of 58% for internal combustion engines. While these efficiencies are not approached in most real world applications, high-temperature fuel cells (solid oxide fuel cells or molten carbonate fuel cells) can theoretically be combined with gas turbines to allow stationary fuel cells to come closer to the theoretical limit. A gas turbine would capture heat from the fuel cell and turn it into mechanical energy to increase the fuel cell's operational efficiency. This solution has been predicted to increase total efficiency to as much as 70%.

In Practice

The tank-to-wheel efficiency of a fuel-cell vehicle is greater than 45% at low loads and shows average values of about 36% when a driving cycle like the NEDC (New European Driving Cycle) is used as test procedure. The comparable NEDC value for a Diesel vehicle is 22%. In 2008 Honda released a demonstration fuel cell electric vehicle (the Honda FCX Clarity) with fuel stack claiming a 60% tank-to-wheel efficiency.

It is also important to take losses due to fuel production, transportation, and storage into account. Fuel cell vehicles running on compressed hydrogen may have a power-plant-to-wheel efficiency of 22% if the hydrogen is stored as high-pressure gas, and 17% if it is stored as liquid hydrogen. Fuel cells cannot store energy like a battery, except as hydrogen, but in some applications, such as stand-alone power plants based on discontinuous sources such as solar or wind power, they are combined with electrolyzers and storage systems to form an energy storage system. Most hydrogen, however, is produced by steam methane reforming, and so most hydrogen production emits carbon dioxide. The overall efficiency (electricity to hydrogen and back to electricity) of such plants (known as *round-trip efficiency*), using pure hydrogen and pure oxygen can be "from 35 up to 50 percent", depending on gas density and other conditions. While a much cheaper lead–acid battery might return about 90%, the electrolyzer/fuel cell system can store indefinite quantities of hydrogen, and is therefore better suited for long-term storage.

Solid-oxide fuel cells produce exothermic heat from the recombination of the oxygen and hydrogen. The ceramic can run as hot as 800 degrees Celsius. This heat can be captured and used to heat water in a micro combined heat and power (m-CHP) application. When the heat is captured, total efficiency can reach 80–90% at the unit, but does not consider production and distribution losses. CHP units are being developed today for the European home market.

Professor Jeremy P. Meyers, in the Electrochemical Society journal *Interface* in 2008, wrote, "While fuel cells are efficient relative to combustion engines, they are not as efficient as batteries, due primarily to the inefficiency of the oxygen reduction reaction (and ... the oxygen evolution reaction, should the hydrogen be formed by electrolysis of water).... [T]hey make the most sense for operation disconnected from the grid, or when fuel can be provided continuously. For applications that require frequent and relatively rapid start-ups ... where zero emissions are a requirement, as in enclosed spaces such as warehouses, and where hydrogen is considered an acceptable reactant, a [PEM fuel cell] is becoming an increasingly attractive choice [if exchanging batteries is inconvenient]". In 2013 military organisations are evaluating fuel cells to significantly reduce the battery weight carried by soldiers.

Applications

***Figure:** Type 212 submarine with fuel cell propulsion of the German Navy in dry dock*

Power

Stationary fuel cells are used for commercial, industrial and residential primary and backup power generation. Fuel cells are very useful as power sources in remote locations, such as spacecraft, remote weather stations, large parks, communications centres, rural locations including research stations, and in certain military applications. A fuel cell system running on hydrogen can be compact and lightweight, and have no major moving parts. Because fuel cells have no moving parts and do not involve combustion, in ideal conditions they can achieve up to 99.9999% reliability. This equates to less than one minute of downtime in a six-year period.

Since fuel cell electrolyzer systems do not store fuel in themselves, but rather rely on external storage units, they can be successfully applied in large-scale energy storage, rural areas being one example. There are many different types of stationary fuel cells so efficiencies vary, but most are between 40% and 60% energy efficient. However, when the fuel cell's waste heat is used to heat a building in a cogeneration system this efficiency can increase to 85%. This is significantly more efficient than traditional coal power plants, which are only about one third energy efficient. Assuming production at scale, fuel cells could save 20–40% on energy costs when used in cogeneration systems. Fuel cells are also much cleaner than traditional power generation; a fuel cell power plant using natural gas as a hydrogen source would create less than one ounce of pollution (other than CO_2) for every 1,000 kW •h produced, compared to 25 pounds of pollutants generated by conventional combustion systems. Fuel Cells also produce 97% less nitrogen oxide emissions than conventional coal-fired power plants.

Coca-Cola, Google, Walmart, Sysco, FedEx, UPS, Ikea, Staples, Whole Foods, Gills Onions, Nestle Waters, Pepperidge Farm, Sierra Nevada Brewery, Super Store Industries, Brigestone-Firestone, Nissan North America, Kimberly-Clark, Michelin and more have installed fuel cells to help meet their power needs. One such pilot program is operating on Stuart Island in Washington State. There the Stuart Island Energy Initiative has built a complete, closed-loop system: Solar panels power an electrolyzer, which makes hydrogen. The hydrogen is stored in a 500-U.S.-gallon (1,900 L) tank at 200 pounds per square inch (1,400 kPa), and runs a ReliOn fuel cell to provide full electric back-up to the off-the-grid residence. Another closed system loop was unveiled in late 2011 in Hempstead, NY.

Fuel cells can be used with low-quality gas from landfills or waste-water treatment plants to generate power and lower methane emissions. A 2.8 MW fuel cell plant in California is said to be the largest of the type.

Cogeneration

Combined heat and power (CHP) fuel cell systems, including Micro combined heat and power (MicroCHP) systems are used to generate both electricity and heat for homes, office building and factories. The system generates constant electric power (selling excess power back to the grid when it is not consumed), and at the same time produces hot air and water from the waste heat. As the result CHP systems have the potential to save primary energy as they can make use of waste heat which is generally rejected by thermal energy conversion systems. A typical capacity range of home fuel cell is 1–3 kW_{el} / 4–8 kW_{th}. CHP systems linked to absorption chillers use their waste heat for refrigeration.

The waste heat from fuel cells can be diverted during the summer directly into the ground providing further cooling while the waste heat during winter can be pumped directly into the building. The University of Minnesota owns the patent rights to this type of system.

Co-generation systems can reach 85% efficiency (40–60% electric + remainder as thermal). Phosphoric-acid fuel cells (PAFC) comprise the largest segment of existing CHP products worldwide and can provide combined efficiencies close to 90%. Molten Carbonate (MCFC) and Solid Oxide Fuel Cells (SOFC) are also used for combined heat and power generation and have electrical energy efficiencies around 60%. Disadvantages of co-generation systems include slow ramping up and down rates, high cost and short lifetime. Also their need to have a hot water storage tank to smooth out the thermal heat production was a serious disadvantage in the domestic market place where space in domestic properties is at a great premium.

Fuel Cell Electric Vehicles (FCEVs)

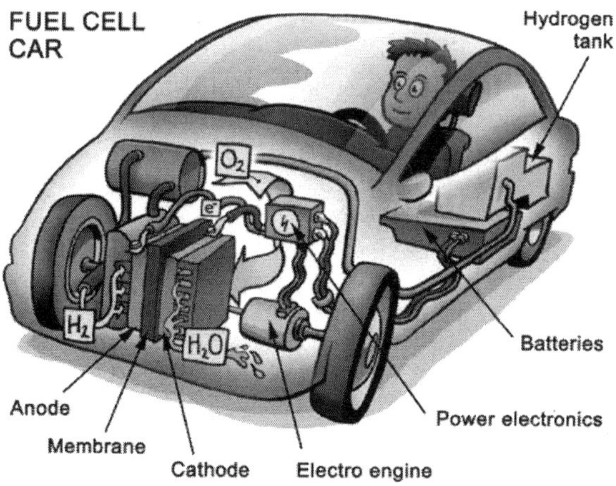

Figure: *Configuration of components in a fuel cell car.*

Automobiles

Although there are currently no Fuel cell vehicles available for commercial sale, over 20 FCEVs prototypes and demonstration cars have been released since 2009. Demonstration models include the Honda FCX Clarity, Toyota FCHV-adv, and Mercedes-Benz F-Cell. As of June 2011 demonstration FCEVs had driven more than 4,800,000 km (3,000,000 mi), with more than 27,000 refuellings.

Demonstration fuel cell vehicles have been produced with "a driving range of more than 400 km (250 mi) between refuelling". They can be refuelled in less than 5 minutes. The U.S. Department of Energy's Fuel Cell Technology Program claims that, as of 2011, fuel cells achieved 53–59% efficiency at one-quarter power and 42–53% vehicle efficiency at full power, and a durability of over 120,000 km (75,000 mi) with less than 10% degradation.

In a Well-to-Wheels simulation analysis, that "did not address the economics and market constraints", General Motors and its partners estimated that per mile travelled, a fuel cell electric vehicle running on compressed gaseous hydrogen produced from natural gas could use about 40% less energy and emit 45% less greenhouse gasses than an internal combustion vehicle. A lead engineer from the Department of Energy whose team is testing fuel cell cars said in 2011 that the potential appeal is that "these are full-function vehicles with no limitations on range or refuelling rate so they are a direct replacement for any vehicle. For instance, if you drive a full sized SUV and pull a boat up into the mountains, you can do that with this technology and you can't with current battery-only vehicles, which are more geared toward city driving."

Some experts believe, however, that fuel cell cars will never become economically competitive with other technologies or that it will take decades for them to become profitable. In July 2011, the chairman and CEO of General Motors, Daniel Akerson, stated that while the cost of hydrogen fuel cell cars is decreasing: "The car is still too expensive and probably won't be practical until the 2020-plus period, I don't know."

In 2012, Lux Research, Inc. issued a report that stated: "The dream of a hydrogen economy ... is no nearer." It concluded that "Capital cost ... will limit adoption to a mere 5.9 GW" by 2030, providing "a nearly insurmountable barrier to adoption, except in niche applications". The analysis concluded that, by 2030, PEM stationary market will reach $1 billion, while the vehicle market, including forklifts, will reach a total of $2 billion. Other analyses cite the lack of an extensive hydrogen infrastructure in the U.S. as an ongoing challenge to Fuel Cell Electric Vehicle commercialization. In 2006, a study for the IEEE showed that for hydrogen produced via electrolysis of water: "Only about 25% of the power generated from wind, water, or sun is converted to practical use." The study further noted that "Electricity obtained from hydrogen fuel cells appears to be four times as expensive as electricity drawn from the electrical transmission grid. ... Because of the high energy losses [hydrogen] cannot compete with electricity." Furthermore, the study found: "Natural gas reforming is not a sustainable solution". "The large amount of energy required to isolate hydrogen from natural compounds (water, natural gas, biomass), package the light gas by compression or liquefaction, transfer the energy carrier to the user, plus the energy lost when it is converted to useful electricity with fuel cells, leaves around 25% for practical use."

Despite this, several major car manufacturers have announced plans to introduce a production model of a fuel cell car in 2015. In 2013, Toyota has stated that it plans to introduce such a vehicle at a price of less than US$100,000.

Mercedes-Benz announced that they would move the scheduled production date of their fuel cell car from 2015 up to 2014, asserting that "The product is ready for the market technically. ... The issue is infrastructure." At the Paris Auto Show in September 2012, Hyundai announced that it plans to begin producing a commercial production fuel cell model (based on the ix35) in December 2012 and hopes to deliver 1,000 of them by 2015. Other manufacturers planning to sell fuel cell electric vehicles commercially by 2016 or earlier include General Motors (2015), Honda (2015 in Japan), and Nissan (2016).

The Obama Administration sought to reduce funding for the development of fuel cell vehicles, concluding that other vehicle technologies will lead to quicker reduction in emissions in a shorter time. Steven Chu, the United States Secretary of Energy, stated in 2009 that hydrogen vehicles "will not be practical over the next 10 to 20 years". In 2012, however, Chu stated that he saw fuel cell cars as more economically feasible as natural gas prices have fallen and hydrogen reforming technologies have improved.

Buses

As of August 2011, there were a total of approximately 100 fuel cell buses deployed around the world. Most buses are produced by UTC Power, Toyota, Ballard, Hydrogenics, and Proton Motor. UTC Buses had accumulated over 970,000 km (600,000 mi) of driving by 2011. Fuel cell buses have a 39–141% higher fuel economy than diesel buses and natural gas buses. Fuel cell buses have been deployed around the world including in Whistler, Canada; San Francisco, United States; Hamburg,

Germany; Shanghai, China; London, England; São Paulo, Brazil; as well as several others.

Figure: *Toyota FCHV-BUS at the Expo 2005.*

The Fuel Cell Bus Club is a global cooperative effort in trial fuel cell buses. Notable Projects Include:

- 12 Fuel cell buses are being deployed in the Oakland and San Francisco Bay area of California.
- Daimler AG, with thirty-six experimental buses powered by Ballard Power Systems fuel cells completed a successful three-year trial, in eleven cities, in January 2007.
- A fleet of Thor buses with UTC Power fuel cells was deployed in California, operated by SunLine Transit Agency.

The first Brazilian hydrogen fuel cell bus prototype in Brazil was deployed in São Paulo. The bus was manufactured in Caxias do Sul and the hydrogen fuel will be produced in São Bernardo do Campo from water through electrolysis. The program, called "*Ônibus Brasileiro a Hidrogênio*" (Brazilian Hydrogen Autobus), includes three additional buses.

Forklifts

A fuel cell forklift (also called a fuel cell lift truck) is a fuel cell powered industrial forklift truck used to lift and transport materials. Most fuel cells used for material handling purposes are powered by PEM fuel cells.

In 2013 there were over 4,000 fuel cell forklifts used in material handling in the USA, of which only 500 received funding from DOE (2012). Fuel cell fleets are operated by a large number of companies, including Sysco Foods, FedEx Freight, GENCO (at Wegmans, Coca-Cola, Kimberly Clark, and Whole Foods), and H-E-B Grocers. Europe demonstrated 30 Fuel cell forklifts with Hylift and extended it with HyLIFT-EUROPE to 200 units, with other projects in France and Austria. Pike Research stated in 2011 that fuel-cell-powered forklifts will be the largest driver of hydrogen fuel demand by 2020.

PEM fuel-cell-powered forklifts provide significant benefits over both petroleum and battery powered forklifts as they produce no local emissions, can work for a full 8-hour shift on a single tank of hydrogen, can be refuelled in 3 minutes and have a lifetime of 8–10 years. Fuel cell-powered forklifts are often used in refrigerated warehouses, as their performance is not degraded by lower temperatures. Many companies do not use petroleum powered forklifts, as these vehicles work indoors where emissions must be controlled and instead are turning to electric forklifts. In design the FC units are often made as drop-in replacements.

Motorcycles and Bicycles

In 2005 a British manufacturer of hydrogen-powered fuel cells, Intelligent Energy (IE), produced the first working hydrogen run motorcycle called the ENV (Emission Neutral Vehicle). The motorcycle holds enough fuel to run for four hours, and to travel 160 km (100 mi) in an urban area, at a top speed of 80 km/h (50 mph). In 2004 Honda developed a fuel-cell motorcycle that utilised the Honda FC Stack.

Other examples of motorbikes and bicycles that use hydrogen fuel cells include the Taiwanese company APFCT's scooter using the fuelling system from Italy's Acta SpA and the Suzuki Burgman scooter with an IE fuel cell that received EU Whole Vehicle Type Approval in 2011. Suzuki Motor Corp. and IE have announced a joint venture to accelerate the commercialization of zero-emission vehicles.

Airplanes

Boeing researchers and industry partners throughout Europe conducted experimental flight tests in February 2008 of a manned airplane powered only by a fuel cell and lightweight batteries. The fuel cell demonstrator airplane, as it was called, used a proton exchange membrane (PEM) fuel cell/lithium-ion battery hybrid system to power an electric motor, which was coupled to a conventional propeller. In 2003, the world's first propeller-driven airplane to be powered entirely by a fuel cell was flown. The fuel cell was a unique FlatStack™ stack design, which allowed the fuel cell to be integrated with the aerodynamic surfaces of the plane.

There have been several fuel-cell-powered unmanned aerial vehicles (UAV). A Horizon fuel cell UAV set the record distance flown for a small UAV in 2007. The military is especially interested in this application because of the low noise, low thermal signature and ability to attain high altitude. In 2009 the Naval Research Laboratory's (NRL's) Ion Tiger utilised a hydrogen-powered fuel cell and flew for 23 hours and 17 minutes. Fuel cells are also being used to provide auxiliary power in aircraft, replacing fossil fuel generators that were previously used to start the engines and power on board electrical needs. Fuel cells can help airplanes reduce CO_2 and other pollutant emissions and noise.

Boats

Figure: *The world's first certified Fuel Cell Boat (HYDRA), in Leipzig/Germany*

The world's first fuel-cell boat HYDRA used an AFC system with 6.5 kW net output. Iceland has committed to converting its vast fishing fleet to use fuel cells to provide auxiliary power by 2015 and, eventually, to provide primary power in its boats. Amsterdam recently introduced its first fuel-cell-powered boat that ferries people around the city's famous and beautiful canals.

Submarines

The Type 212 submarines of the German and Italian navies use fuel cells to remain submerged for weeks without the need to surface.

The U212A is a non-nuclear submarine developed by German naval shipyard Howaldtswerke Deutsche Werft. The system consists of nine

PEM fuel cells, providing between 30 kW and 50 kW each. The ship is silent giving it an advantage in the detection of other submarines.

Portable Power Systems

Portable power systems that use fuel cells can be used in the leisure sector (i.e. RV's, Cabins, Marine), the industrial sector (i.e. power for remote locations including gas/oil wellsites, communication towers, security, weather stations etc.), and in the military sector. SFC Energy is a German manufacturer of direct methanol fuel cells for a variety of portable power systems. Ensol Systems Inc. is an integrator of portable power systems, using the SFC Energy DMFC.

Other Applications

- Providing power for base stations or cell sites
- Distributed generation
- Emergency power systems are a type of fuel cell system, which may include lighting, generators and other apparatus, to provide backup resources in a crisis or when regular systems fail. They find uses in a wide variety of settings from residential homes to hospitals, scientific laboratories, data centres,
- telecommunication equipment and modern naval ships.
- An uninterrupted power supply (UPS) provides emergency power and, depending on the topology, provide line regulation as well to connected equipment by supplying power from a separate source when utility power is not available. Unlike a standby generator, it can provide instant protection from a momentary power interruption.
- Base load power plants
- Solar Hydrogen Fuel Cell Water Heating
- Hybrid vehicles, pairing the fuel cell with either an ICE or a battery.
- Notebook computers for applications where AC charging may not be readily available.
- Portable charging docks for small electronics (e.g. a belt clip that charges your cell phone or PDA).
- Smartphones, laptops and tablets.
- Small heating appliances
- Food preservation, achieved by exhausting the oxygen and automatically maintaining oxygen exhaustion in a shipping container, containing, for example, fresh fish.

- Breathalyzers, where the amount of voltage generated by a fuel cell is used to determine the concentration of fuel (alcohol) in the sample.
- Carbon monoxide detector, electrochemical sensor.

Fuelling Stations

There were over 85 hydrogen refuelling stations in the U.S. in 2010.

As of June 2012 California had 23 hydrogen refuelling stations in operation. Honda announced plans in March 2011 to open the first station that would generate hydrogen through solar-powered renewable electrolysis. South Carolina also has two hydrogen fuelling stations, in Aiken and Columbia, SC. The University of South Carolina, a founding member of the South Carolina Hydrogen & Fuel Cell Alliance, received 12.5 million dollars from the United States Department of Energy for its Future Fuels Program.

The first public hydrogen refuelling station in Iceland was opened in Reykjavík in 2003. This station serves three buses built by DaimlerChrysler that are in service in the public transport net of Reykjavík. The station produces the hydrogen it needs by itself, with an electrolyzing unit (produced by Norsk Hydro), and does not need refilling: all that enters is electricity and water. Royal Dutch Shell is also a partner in the project. The station has no roof, in order to allow any leaked hydrogen to escape to the atmosphere.

The current 14 stations nationwide in Germany are planned to be expanded to 50 by 2015 through its public private partnership Now GMBH. Japan also has a hydrogen highway, as part of the Japan hydrogen fuel cell project. Twelve hydrogen fuelling stations have been built in 11 cities in Japan, and additional hydrogen stations could potentially be operational by 2015. Canada, Sweden and Norway also have hydrogen highways being implemented.

Markets and Economics

In 2012, fuel cell industry revenues exceeded $1 billion market value worldwide, with Asian pacific countries shipping more than 3/4 of the fuel cell systems worldwide. However, as of October 2013, no public company in the industry had yet become profitable. There were 140,000 fuel cell stacks shipped globally in 2010, up from 11 thousand shipments in 2007, and from 2011 to 2012 worldwide fuel cell shipments had an annual growth rate of 85%. Tanaka Kikinzoku Kogyo K.K. expanded its production facilities for fuel cell catalysts in 2013 to meet anticipated demand as the Japanese ENE Farm scheme expects to install 50,000 units in 2013 and the company is experiencing rapid market growth.

Approximately 50% of fuel cell shipments in 2010 were stationary fuel cells, up from about a third in 2009, and the four dominant producers in the Fuel Cell Industry were the United States, Germany, Japan and South Korea. The Department of Energy Solid State Energy Conversion Alliance found that, as of January 2011, stationary fuel cells generated power at approximately \$724 to \$775 per kilowatt installed. In 2011, Bloom Energy, a major fuel cell supplier, said that its fuel cells generated power at 9–11 cents per kilowatt-hour, including the price of fuel, maintenance, and hardware.

Industry groups predict that there are sufficient platinum resources for future demand, and in 2007, research at Brookhaven National Laboratory suggested that platinum could be replaced by a gold-palladium coating, which may be less susceptible to poisoning and thereby improve fuel cell lifetime. Another method would use iron and sulphur instead of platinum. This would lower the cost of a fuel cell (as the platinum in a regular fuel cell costs around US\$1,500, and the same amount of iron costs only around US\$1.50). The concept was being developed by a coalition of the John Innes Centre and the University of Milan-Bicocca. PEDOT cathodes are immune to monoxide poisoning.

Nanoelectronics

Nanoelectronics refer to the use of nanotechnology in electronic components. The term covers a diverse set of devices and materials, with the common characteristic that they are so small that inter-atomic interactions and quantum mechanical properties need to be studied extensively. Some of these candidates include: hybrid molecular/semiconductor electronics, one-dimensional nanotubes/nanowires, or advanced molecular electronics. Recent silicon CMOS technology generations, such as the 22nm node, are already within this regime. Nanoelectronics are sometimes considered as disruptive technology because present candidates are significantly different from traditional transistors.

In 1965 Gordon Moore observed that silicon transistors were undergoing a continual process of scaling downward, an observation which was later codified as Moore's law. Since his observation transistor minimum feature sizes have decreased from 10 micrometres to the 28-22 nm range in 2011. The field of nanoelectronics aims to enable the continued realisation of this law by using new methods and materials to build electronic devices with feature sizes on the nanoscale.

The volume of an object decreases as the third power of its linear dimensions, but the surface area only decreases as its second power. This somewhat subtle and unavoidable principle has huge ramifications. For example the power of a drill (or any other machine) is proportional to the volume, while the friction of the drill's bearings and gears is proportional to their surface area. For a normal-sized drill, the power of the device is enough to handily overcome any friction. However, scaling its length down by a factor of 1000, for example, decreases its power by 1000^3 (a factor of a billion) while reducing the friction by only 1000^2 (a factor of "only" a million). Proportionally it has 1000 times less power per unit friction than the original drill. If the original friction-to-power ratio was, say, 1%, that implies the smaller drill will have 10 times as much friction as power; the drill is useless.

For this reason, while super-miniature electronic integrated circuits are fully functional, the same technology cannot be used to make working mechanical devices beyond the scales where frictional forces start to exceed the available power. So even though you may see microphotographs of delicately etched silicon gears, such devices are currently little more than curiosities with limited real world applications, for example, in moving mirrors and shutters. Surface tension increases in much the same way, thus magnifying the tendency for very small objects to stick together. This could possibly make any kind of "micro factory" impractical: even if robotic arms and hands could be scaled down, anything they pick up will tend to be impossible to put down. The above being said, molecular evolution has resulted in working cilia, flagella, muscle fibres and rotary motors in aqueous environments, all on the nanoscale. These machines exploit the increased frictional forces found at the micro or nanoscale. Unlike a paddle or a propeller which depends on normal frictional forces (the frictional forces perpendicular to the surface) to achieve propulsion, cilia develop motion from the exaggerated drag or laminar forces (frictional forces parallel to the surface) present at micro and nano dimensions. To build meaningful "machines" at the nanoscale, the relevant forces need to be considered. We are faced with the development and design of intrinsically pertinent machines rather than the simple reproductions of macroscopic ones.

All scaling issues therefore need to be assessed thoroughly when evaluating nanotechnology for practical applications.

Approaches to Nanoelectronics

Nanofabrication: For example, single electron transistors, which involve transistor operation based on a single electron. Nanoelectromechanical systems also fall under this category. Nanofabrication can be used to construct ultradense parallel arrays of nanowires, as an alternative to synthesizing nanowires individually.

Nanomaterials Electronics

Besides being small and allowing more transistors to be packed into a single chip, the uniform and symmetrical structure of nanotubes allows a higher electron mobility (faster electron movement in the material), a higher dielectric constant (faster frequency), and a symmetrical electron/hole characteristic.

Molecular Electronics

Single molecule devices are another possibility. These schemes would make heavy use of molecular self-assembly, designing the device components to construct a larger structure or even a complete system on their own. This can be very useful for reconfigurable computing, and may even completely replace present FPGA technology.

Molecular electronics is a new technology which is still in its infancy, but also brings hope for truly atomic scale electronic systems in the future. One of the more promising applications of molecular electronics was proposed by the IBM researcher Ari Aviram and the theoretical chemist Mark Ratner in their 1974 and 1988 papers *Molecules for Memory, Logic and Amplification.*

This is one of many possible ways in which a molecular level diode / transistor might be synthesized by organic chemistry. A model system was proposed with a spiro carbon structure giving a molecular diode about half a nanometre across which could be connected by polythiophene molecular wires. Theoretical calculations showed the design to be sound in principle and there is still hope that such a system can be made to work.

Energy Production

Research is ongoing to use nanowires and other nanostructured materials with the hope to create cheaper and more efficient solar cells than are possible with conventional planar silicon solar cells. It is believed that the invention of more efficient solar energy would have a great effect on satisfying global energy needs.

There is also research into energy production for devices that would operate *in vivo*, called bio-nano generators. A bio-nano generator is a nanoscale electrochemical device, like a fuel cell or galvanic cell, but drawing power from blood glucose in a living body, much the same as how the body generates energy from food. To achieve the effect, an enzyme is used that is capable of stripping glucose of its electrons, freeing them for use in electrical devices. The average person's body could, theoretically, generate 100 watts of electricity (about 2000 food calories per day) using a bio-nano generator. However, this estimate is only true if all food was converted to electricity, and the human body needs some energy consistently, so possible power generated is likely much lower. The electricity generated by such a device could power devices embedded in the body (such as pacemakers), or sugar-fed nanorobots. Much of the research done on bio-nano generators is still experimental, with Panasonic's Nanotechnology Research Laboratory among those at the forefront.

Nanocatalysis for Iron Catalyzed Fischer Tropsch

The Fischer–Tropsch process is a collection of chemical reactions that converts a mixture of carbon monoxide and hydrogen into liquid hydrocarbons. It was first developed by Franz Fischer and Hans Tropsch at the "Kaiser-Wilhelm-Institut für Kohlenforschung" in Mülheim an der Ruhr, Germany in 1925. The process, a key component of gas to liquids technology, produces a synthetic lubrication oil and synthetic fuel, typically from coal, natural gas, or biomass. The Fischer–Tropsch process has received intermittent attention as a source of low-sulfur diesel fuel and to address the supply or cost of petroleum-derived hydrocarbons.

The Fischer–Tropsch process involves a series of chemical reactions that produce a variety of hydrocarbons, ideally having the formula ($C_nH_{(2n+2)}$). The more useful reactions produce alkanes as follows:

$$(2n + 1)\ H_2 + n\ CO \rightarrow C_nH_{(2n+2)} + n\ H_2O$$

where n is typically 10-20. The formation of methane (n = 1) is unwanted. Most of the alkanes produced tend to be straight-chain, suitable as diesel fuel. In addition to alkane formation, competing reactions give small amounts of alkenes, as well as alcohols and other oxygenated hydrocarbons.

Fischer Tropsch Intermediates and Elemental Reactions

Converting a mixture of H_2 and CO into aliphatic products obviously should be a multi-step reaction with several sorts of intermediates. The growth of the hydrocarbon chain may be visualized as involving a repeated sequence in which hydrogen atoms are added to carbon and oxygen, the C/O-bond is split and a new C/C-bond is formed. For one CH_2-group produced $CO + 2H_2 \rightarrow (CH_2) + H_2O$, several reactions are necessary • Associative adsorption of CO • Splitting of the C/O-bond • Dissociative adsorption of $2H_2$ • Transfer of 2H to the oxygen to yield H_2O • Desorption of H_2O • Transfer of 2H to the carbon to yield CH_2.

The conversion of CO to alkanes involves hydrogenation of CO, the hydrogenolysis (cleavage with H_2) of C-O bonds, and the formation of C-C bonds. Such reactions are assumed to proceed via initial formation of surface-bound metal carbonyls. The CO ligand is speculated to undergo dissociation, possibly into oxide and carbide ligands. Other potential intermediates are various C-1 fragments including formyl (CHO), hydroxycarbene (HCOH), hydroxymethyl (CH_2OH), methyl (CH_3), methylene (CH_2), methylidyne (CH), and hydroxymethylidyne (COH). Furthermore, and critical to the production of liquid fuels, are reactions that form C-C bonds, such as migratory insertion. Many related stoichiometric reactions have been simulated on discrete metal clusters, but homogeneous Fischer–Tropsch catalysts are poorly developed and of no commercial importance.

Addition of isotopically labelled alcohol to the feed stream results in incorporation of alcohols into product. This observation establishes the facility of C-O bond scission. Using ^{14}C-labelled ethyene and propene over cobalt catalysts results in incorporation of these olefins into the growing chain. Chain growth reaction thus appears to involve both 'olefin insertion' as well as 'CO-insertion'.

Feedstocks: Gasification

Fischer–Tropsch plants associated with coal or related solid feedstocks (sources of carbon) must first convert the solid fuel into gaseous reactants, *i.e.*, CO, H_2, and alkanes. This conversion is called gasification and the product is called synthesis gas ("Syn gas"). Synthesis gas obtained from coal gasification tends to have a H_2/CO ratio of ~0.7 compared to the ideal ratio of ~2. This ratio is adjusted via the water-gas shift reaction. Coal-based Fischer–Tropsch plants produce varying amounts of CO_2, depending upon the energy source of the gasification process. However, most coal-based plants rely on the feed coal to supply all the energy requirements of the Fischer–Tropsch process.

Feedstocks: GTL

Carbon monoxide for FT catalysis is derived from hydrocarbons. In gas to liquids (GTL) technology, the hydrocarbons are low molecular weight materials that often would be discarded or flared. GTL is economically viable when the gas price is relatively cheap on an energy equivalency basis to oil. Stranded gas provides relatively cheap gas. GTL is viable provided gas remains relatively cheaper than oil.

Several reactions are required to obtain the gaseous reactants required for Fischer–Tropsch catalysis. First, reactant gases entering a Fischer–Tropsch reactor must be desulfurized. Otherwise, sulfur-containing impurities deactivate ("poison") the catalysts required for Fischer–Tropsch reactions.

Several reactions are employed to adjust the H_2/CO ratio. Most important is the water gas shift reaction, which provides a source of hydrogen at the expense of carbon monoxide:

$$H_2O + CO \rightarrow H_2 + CO_2$$

For Fischer–Tropsch plants that use methane as the feedstock, another important reaction is steam reforming, which converts the methane into CO and H_2:

$$H_2O + CH_4 \rightarrow CO + 3\ H_2$$

Process Conditions

Generally, the Fischer–Tropsch process is operated in the temperature range of 150–300 °C (302–572 °F). Higher temperatures lead to faster reactions and higher conversion rates but also tend to favour methane production. For this reason, the temperature is usually maintained at the low to middle part of the range. Increasing the pressure leads to higher conversion rates and also favours formation of long-chained alkanes, both of which are desirable. Typical pressures range from one to several tens of atmospheres. Even higher pressures would be favourable, but the benefits may not justify the additional costs of high-pressure equipment, and higher pressures can lead to catalyst deactivation via coke formation.

A variety of synthesis-gas compositions can be used. For cobalt-based catalysts the optimal H_2:CO ratio is around 1.8–2.1. Iron-based catalysts promote the water-gas-shift reaction and thus can tolerate lower ratios. This reactivity can be important for synthesis gas derived from coal or biomass, which tend to have relatively low H_2:CO ratios (<1).

Design of the Fischer-Tropsch Process Reactor

Efficient removal of heat from the reactor is basic need of the Fischer-Tropsch reactors since these reactions are characterized by high exothermicity. Four types of reactors are discussed:

Multi tubular fixed bed reactor: This type of reactor contains number of tubes with small diameter. These tubes contains catalyst and are surrounded by boiling water which removes the heat of reaction. Fixed Bed Reactor is suitable to operate at low temperatures and has upper temperature limit of 530 K. Excess temperature leads to carbon deposition and hence blockage of the reactor. Since large amounts of products formed are in liquid state this type reactor can also be termed as trickle flow reactor system.

Entrained flow reactor: Important requirement of the reactor for Fischer Tropsch process is to remove the heat of the reaction. This type of reactor contains two banks of heat exchangers which removes heat, remaining is removed by products and recycle of the system. The formation of heavy waxes should be avoided since this condenses on the catalyst and forms agglomeration. This leads to fluidization. Hence risers are operated over 570 K.

Slurry Reactors: Heat removal is done by internal cooling coils. Synthesis gas is bubbled through the waxy products and finely divided catalyst which is suspended in the liquid medium. This also provides agitation of the contents of the reactor. Catalyst particle size reduces diffusional heat and mass transfer limitations. Lower temperature in reactor leads to more viscous product and higher (> 570 K) gives undesirable products spectrum. However separation of the product from catalyst is a problem.

Fluid-bed- and circulating catalyst(riser) reactors: These are used for high temperature Fischer Tropsch synthesis (nearly 340°C) to produce low molecular weight olefinic hydrocarbons on alkalised fused iron catalysts. The fluid bed technology (as adapted from catalytic cracking of heavy petroleum distillates) was introduced by Hydrocarbon Research in the years 1946–1950 and named 'Hydrocol' process. A large scale Fischer Tropsch Hydrocol plant (3,50,000 tons per annum) operated during the years 1951–1957 in Brownsville, Texas. Due to technical problems and lacking economy at increasing petroleum availability this development was discontinued.

Fluid bed Fischer Tropsch synthesis has recently been very successfully reinvestigated by Sasol. One reactor with a capacity of 5,00,000 tons per annum is now in operation and even larger ones are being built (nearly 8,50,000 tons per annum). The process is now used for mainly olefins C_2, C^7 production. This new development can be regarded as an important progress in Fischer Tropsch technology. A high temperature process with a circulating iron catalyst ('Circulating fluid bed', 'riser reactor', 'entrained catalyst process') was introduced by the Kellogg Company and a respective plant built at Sasol in 1956. It was improved by Sasol for successful operation. At Secunda, South Africa, Sasol has operated 16 advanced reactors of this type with a capacity of Approx. 3,30,000 tons per annum each. Now the circulating catalyst process is being replaced by the superior Sasol advanced fluid bed technology. Early experiments with cobalt-catalyst particles suspended in oil have been performed by Fischer. The bubble column reactor with a powdered iron slurry catalyst and a CO-rich syngas was particularly developed to pilot plant scale by Kölbel at the Rheinpreuben Company in 1953. Recently (since 1990) low temperature Fischer Tropsch slurry processes are under investigation for the use of iron and cobalt catalysts, particularly for the production of a hydrocarbon wax, to be used as such, or to be hydrocracked and isomerised to mainly Diesel fuel by Exxon and Sasol.

Today slurry phase (bubble column) low temperature Fischer Tropsch synthesis is regarded by many authors as the most efficient process for Fischer Tropsch clean Diesel production. This Fischer Tropsch technology is also under development by the Statoil Company (Norway) for use on a vessel to convert associated gas at offshore oil fields into a hydrocarbon liquid.

Product Distribution

In general the product distribution of hydrocarbons formed during the Fischer–Tropsch process follows an Anderson–Schulz–Flory distribution, which can be expressed as:

$$W_n/n = (1 - \alpha)^2\alpha^{n-1}$$

where W_n is the weight fraction of hydrocarbons containing n carbon atoms. α is the chain growth probability or the probability that a molecule will continue reacting to form a longer chain. In general, α is largely determined by the catalyst and the specific process conditions.

Examination of the above equation reveals that methane will always be the largest single product so long as alpha is less than 0.5; however, by increasing α close to one, the total amount of methane formed can be minimized compared to the sum of all of the various long-chained products. Increasing α increases the formation of long-chained hydrocarbons. The very long-chained hydrocarbons are waxes, which are solid at room temperature. Therefore, for production of liquid transportation fuels it may be necessary to crack some of the Fischer–Tropsch products. In order to avoid this, some researchers have proposed using zeolites or other catalyst substrates with fixed sized pores that can restrict the formation of hydrocarbons longer than some characteristic size (usually $n<10$). This way they can drive the reaction so as to minimize methane formation without producing lots of long-chained hydrocarbons. Such efforts have met with only limited success.

Catalysts

A variety of catalysts can be used for the Fischer–Tropsch process, but the most common are the transition metals cobalt, iron, and ruthenium. Nickel can also be used, but tends to favour methane formation ("methanation").

Cobalt-based catalysts are highly active, although iron may be more suitable for certain applications. Cobalt catalysts are more active for Fischer–Tropsch synthesis when the feedstock is natural gas. Natural gas has a high hydrogen to carbon ratio, so the water-gas-shift is not needed for cobalt catalysts. Iron catalysts are preferred for lower quality feedstocks such as coal or biomass. Synthesis gases derived from these hydrogen-poor feedstockes has a low-hydrogen-content and require the water-gas-shift reaction. Unlike the other metals used for this process (Co, Ni, Ru), which remain in the metallic state during synthesis, iron catalysts tend to form a number of phases, including various oxides and carbides during the reaction. Control of these phase transformations can be important in maintaining catalytic activity and preventing breakdown of the catalyst particles.

In addition to the active metal the catalysts typically contain a number of "promoters," including potassium and copper. Group 1 alkali metals, including potassium, are a poison for cobalt catalysts but are promoters for iron catalysts. Catalysts are supported on high-surface-area binders/supports such as silica, alumina, or zeolites. Promotors also have an important influence on activity. Alkali metal oxides and copper are common promotors, but the formulation depends on the primary metal, iron vs cobalt. Alkali oxides on cobalt catalysts generally cause activity to drop severely even with very low alkali loadings. C5+ and CO2 selectivity increase while methane and C2-C4 selectivity decrease. In addition, the olefin to parafin ratio increases.

Fischer–Tropsch catalysts are sensitive to poisoning by sulfur-containing compounds. Cobalt-based catalysts are more sensitive than for their iron counterparts. Iron as catalyst:Fischer Tropsch iron catalysts need alkali promotion to attain high activity and stability (e.g. 0.5 wt.% K2O). Addition of Cu for reduction promotion, addition of SiO2, Al2O3 for structural promotion and maybe some manganese can be applied for selectivity control (e.g. high olefinicity). The working catalyst is only obtained when – after reduction with hydrogen– in the initial period of synthesis several iron carbide phases and elemental carbon are formed whereas iron oxides are still present in addition to some metallic iron. With iron catalysts two directions of selectivity have been pursued. One direction has aimed at a low molecular weight olefinic hydrocarbon mixture to be produced in an entrained phase or fluid bed process (Sasol Synthol process). Due to the relatively high reaction temperature (approx. 340°C), the average molecular weight of the product is so low that no liquid product phase occurs under reaction conditions. The catalyst particles moving around in the reactor are small (particle diameter 100 mm) and carbon deposition on the catalyst does not disturb reactor operation.

Thus a low catalyst porosity with small pore diameters as obtained from fused magnetite (plus promotors) after reduction with hydrogen is appropriate. For maximising the overall gasoline yield the olefins C_3, C_4 have been oligomerised at Sasol. However, recovering the olefins for use as chemicals in e.g. polymerization processes is advantageous today. The second direction of iron catalyst development has aimed at highest catalyst activity to be used at low reaction temperature where most of the hydrocarbon product is in the liquid phase under reaction conditions. Typically, such catalysts are obtained through precipitation from nitrate solutions. A high content of a carrier provides mechanical strength and wide pores for easy mass transfer of the reactants in the liquid product filling the pores. The main product fraction then is a paraffin wax, which is refined to marketable wax materials at Sasol, however, also can be very selectively hydrocracked to a high quality Diesel fuel. Thus iron catalysts are very flexible.

Ruthenium as a Fischer Tropsch catalyst:It is most active working at the lowest reaction temperature. It produces the highest molecular weight hydrocarbons ('polymethylene synthesis' Pichler and Buffleb performing thus the chain growth reaction in the cleanest mode; it acts as a Fischer Tropsch catalyst as the pure metal, without any promotors, thus providing the simplest catalytic system of Fischer Tropsch synthesis, where mechanistic conclusions should be the easiest – much easier than e.g. with iron as the catalyst.

Like with nickel, the selectivity changes to mainly methane at elevated temperature. Its high price and limited world resources exclude industrial application. Systematic Fischer Tropsch studies with ruthenium catalysts should contribute substantially to the further exploration of the fundamentals of Fischer Tropsch synthesis. There is an interesting question to consider: what features have the metals nickel, iron, cobalt and ruthenium in common to let them be – and only them – Fischer Tropsch catalysts, converting the CO/H_2 mixture to aliphatic (long chain) hydrocarbons in a 'one step reaction'. The term 'one step reaction' means that reaction intermediates are not desorbed from the catalyst surface. In particular, it is amazing that the much carbided alkalized iron catalyst gives this reaction similarly as the just metallic ruthenium catalyst. It will be seen in the next section that the kinetic principle of 'selective inhibition' might be the common feature which applies, in spite of differences in catalyst composition, reaction intermediates, steps of reaction and corresponding kinetic schemes.

LTFT and HTFT

High-temperature Fischer–Tropsch (or HTFT) is operated at temperatures of 330°C–350°C and uses an iron-based catalyst. This process was used extensively by Sasol in their coal-to-liquid plants (CTL). Low-Temperature Fischer–Tropsch (LTFT) is operated at lower temperatures and uses a cobalt based catalyst. This process is best known for being used in the first integrated Gas-to-Liquid (GTL) plant operated and built by Shell in Bintulu, Malaysia.

History

Since the invention of the original process by Fischer and Tropsch, working at the Kaiser-Wilhelm-Institut for Chemistry in the 1920s, many refinements and adjustments were made. Fischer and Tropsch filed a number of patents, *e.g.*, U.S. Patent 1,746,464, applied 1926, published 1930. It was commercialized by Brabag in Germany in 1936. Being petroleum-poor but coal-rich, Germany used the Fischer–Tropsch process during World War II to produce *ersatz* (replacement) fuels. Fischer–Tropsch production accounted for an estimated 9% of German war production of fuels and 25% of the automobile fuel.

The United States Bureau of Mines, in a program initiated by the Synthetic Liquid Fuels Act, employed seven Operation Paperclip synthetic fuel scientists in a Fischer–Tropsch plant in Louisiana, Missouri in 1946.

In Britain, Alfred August Aicher obtained several patents for improvements to the process in the 1930s and 1940s. Aicher's company was named *Synthetic Oils Ltd.* (Not related to a company of the same name in Canada)

Commercialization

The Fischer–Tropsch process has been applied in large-scale gas–liquids and coal–liquid facilities such as Shell's Pearl GTL facility in Ras Laffan, Qatar. Such large facilities are susceptible to high capital costs, high operation and maintenance costs, the uncertain and volatile price of crude oil, and environmental concerns. In particular, the use of natural gas as a feedstock becomes practical only with use of "stranded gas", *i.e.*, sources of natural gas far from major cities which are impractical to exploit with conventional gas pipelines and LNG technology; otherwise, the direct sale of natural gas to consumers would become much more profitable. Several companies are developing the process to enable practical exploitation of so-called stranded gas reserves.

***Figure:** Fluidized bed gasification with FT-pilot in Güssing, Burgenland, Austria*

Conventional FT reactors have been optimized for massive coal-to-liquids and gas–liquid facilities such as Shell's Pearl GTL facility. These slurry bed and fixed-bed reactors are much larger than the sizes needed for biofuel facilities or for smaller-scale natural-gas fields. The use of microchannel reactors scales down the size of the reaction hardware and overcomes the heat and mass transport problems associated with conventional FT technology. Enhanced heat transfer inside the microchannels reactor allows for optimal temperature control, which maximizes catalyst activity and life. While no smaller scale plant is currently in commercial operation, indications show capital costs, operating costs and size could all be reduced relative to conventional FT facilities. An order has reportedly been placed for a 1400-bbl/day modular GTL plant using the technology of a company called Velocys.

In Australia, Linc Energy commenced construction in 1999 of the world's first gas–liquid plant operating on synthesis gas produced by underground coal gasification. The GTL plant uses the F-T process, and produced liquids in 2008.

Sasol

Figure: *A SASOL garage in Gauteng*

The largest scale implementation of Fischer–Tropsch technology are in a series of plants operated by Sasol in South Africa, a country with large coal reserves, but little oil. The first commercial plant opening in 1952, 40 miles south of Johannesburg. Sasol uses coal and now natural gas as feedstocks and produces a variety of synthetic petroleum products, including most of the country's diesel fuel.

In December, 2012 Sasol announced plans to build a 96,000 barrels a day plant in Westlake, Louisiana using natural gas from tight shale formations in Louisiana and Texas as feedstock. Costs are estimated to be between 11 and 12 billion dollars with $2 billion in tax relief being contributed by the state of Louisiana. The planned complex will include a refinery and a chemical plant.

PetroSA

PetroSA, a South African company which, in a joint venture, won project innovation of the year award at the Petroleum Economist Awards in 2008, has the world's largest Gas to Liquids complexes at Mossel Bay in South Africa. The refinery is a 36,000 barrels a day plant that completed semi-commercial demonstration in 2011, paving the way to begin commercial preparation. The technology can be used to convert natural gas, biomass or coal into synthetic fuels.

Shell Middle Distillate Synthesis

One of the largest implementations of Fischer–Tropsch technology is in Bintulu, Malaysia. This Shell facility converts natural gas into low-sulfur Diesel fuels and food-grade wax. The scale is 12,000 barrels per day (1,900 m^3/d).

Ras Laffan, Qatar

The new LTFT facility Pearl GTL which began operation in 2011 at Ras Laffan, Qatar, uses cobalt catalysts at 230°C, converting natural gas to petroleum liquids at a rate of 140,000 barrels per day (22,000 m^3/d), with additional production of 120,000 barrels (19,000 m^3) of oil equivalent in natural gas liquids and ethane. The first GTL plant in Ras Laffan was commissioned in 2007 and is called Oryx GTL and has a capacity of 34 000 bbl/day. The plant utilises the Sasol slurry phase distillate process which uses a cobalt catalyst. Oryx GTL is a joint venture between Qatar Petroleum and Sasol.

UPM (Finland)

In October 2006, Finnish paper and pulp manufacturer UPM announced its plans to produce biodiesel by the Fischer–Tropsch process alongside the manufacturing processes at its European paper and pulp plants, using waste biomass resulting from paper and pulp manufacturing processes as source material.

Rentech

A demonstration-scale Fischer–Tropsch plant is owned and operated by Rentech, Inc., in partnership with ClearFuels, a company specialising in biomass gasification. Located in Commerce City, Colorado, the facility produces about 10 barrels per day (1.6 m^3/d) of fuels from natural gas. Commercial-scale facilities are planned for Rialto, California; Natchez, Mississippi; Port St. Joe, Florida; and White River, Ontario. Rentech closed down their pilot plant in 2013, and does not appear to be continuing work on their FT process and the proposed commercial facilities.

Other

In the United States, some coal-producing states have invested in Fischer–Tropsch plants. In Pennsylvania, Waste Management and

Processors, Inc. was funded by the state to implement Fischer–Tropsch technology licensed from Shell and Sasol to convert so-called waste coal (leftovers from the mining process) into low-sulfur diesel fuel.

Research Developments

Choren Industries has built an Fischer–Tropsch plant in Germany that converts biomass to syngas and fuels using the Shell Fischer–Tropsch process structure. The company went bankrupt in 2011 due to impracticalities in the process.

Biomass gasification (BG) and Fischer–Tropsch (FT) synthesis can in principle be combined to produce renewable transportation fuels (biofuels).

Carbon Dioxide Reuse

Carbon dioxide is not a typical feedstock for F-T catalysis. Hydrogen and carbon dioxide react over a cobalt-based catalyst, producing methane. With iron-based catalysts short-chain, unsaturated hydrocarbons are also produced. Upon introduction to the catalyst's support, ceria functions as a reverse water gas shift catalyst, further increasing the yield of the reaction. The short chain hydrocarbons were upgraded to liquid fuels over solid acid catalysts, such as zeolites.

Process Efficiency

Using conventional FT technology the process ranges in carbon efficiency from 25 to 50 percent and a thermal efficiency of about 50% for CTL facilities idealised at 60% with GTL facilities at about 60% efficiency idealised to 80% efficiency.

3

Nanotechnology for Energy Production

Nanotechnologies provide the potential to enhance energy efficiency across all branches of industry and to economically leverage renewable energy production through new technological solutions and optimized production technologies. Nanotechnology innovations could impact each part of the value-added chain in the energy sector:

Energy Sources

Nanotechnologies provide essential improvement potentials for the development of both conventional energy sources (fossil and nuclear fuels) and renewable energy sources like geothermal energy, sun, wind, water, tides or biomass. Nano-coated, wear resistant drill probes, for example, allow the optimization of lifespan and efficiency of systems for the development of oil and natural gas deposits or geothermal energy and thus the saving of costs. Further examples are high-duty nanomaterials for lighter and more rugged rotor blades of wind and tidepower plants as well as wear and corrosion protection layers for mechanically stressed components (bearings, gear boxes, etc.). Nanotechnologies will play a decisive role in particular in the intensified use of solar energy through photovoltaic systems. In case of conventional crystalline silicon solar cells, for instance, increases in efficiency are achievable by antireflection layers for higher light yield.

First and foremost, however, it will be the further development of alternative cell types, such as thin-layer solar cells (among others of silicon or other material systems like copper/indium/selenium), dye solar cells or polymer solar cells, which will predominantly profit from nanotechnologies. Polymer solar cells are said to have high potential especially regarding the supply of portable electronic devices, due to the reasonably-priced materials and production methods as well as the flexible design. Medium-term development targets are an efficiency of approx. 10% and a lifespan of several years. Here, for example, nanotechnologies could contribute to the optimization of the layer design and the morphology of organic semiconductor mixtures in component structures. In the long run, the utilisation of nanostructures, like quantum dots and wires, could allow for solar cell efficiencies of over 60%.

Energy Conversion

The conversion of primary energy sources into electricity, heat and kinetic energy requires utmost efficiency. Efficiency increases, especially in fossil-fired gas and steam power plants, could help avoid considerable amounts of carbon dioxide emissions.

Higher power plant efficiencies, however, require higher operating temperatures and thus heat-resistant turbine materials. Improvements are possible, for example, through nano-scale heat and corrosion protection layers for turbine blades in power plants or aircraft engines to enhance the efficiency through increased operating temperatures or the application of lightweight construction materials (e.g. titanium aluminides).

Nano-optimized membranes can extend the scope of possibilities for separation and climate-neutral storage of carbon dioxide for power generation in coal-fired power plants, in order to render this important method of power generation environmentally friendlier in the long run. The energy yield from the conversion of chemical energy through fuel cells can be stepped up by nano-structured electrodes, catalysts and membranes, which results in economic application possibilities in automobiles, buildings and the operation of mobile electronics.

Thermoelectric energy conversion seems to be comparably promising. Nano-structured semiconductors with optimized boundary layer design contribute to increases in efficiency that could pave the way for a broad application in the utilisation of waste heat, for example in automobiles, or even of human body heat for portable electronics in textiles.

Energy Distribution

Regarding the reduction of energy losses in current transmission, hope exists that the extraordinary electric conductivity of nanomaterials like carbon nanotubes can be utilised for application in electric cables and power lines. Furthermore, there are nanotechnological approaches for the optimization of superconductive materials for lossless current conduction. In the long run, options are given for wireless energy transport, e.g. through laser, microwaves or electromagnetic resonance. Future power distribution will require power systems providing dynamic load and failure management, demand-driven energy supply with flexible price mechanisms as well as the possibility of feeding through a number of decentralized renewable energy sources.

Nanotechnologies could contribute decisively to the realisation of this vision, inter alia, through nano-sensory devices and power-electronical components able to cope with the extremely complex control and monitoring of such grids.

Energy Storage

The utilisation of nanotechnologies for the enhancement of electrical energy stores like batteries and super-capacitors turns out to be downright promising. Due to the high cell voltage and the outstanding energy and power density, the lithium-ion technology is regarded as the most promising variant of electrical energy storage. Nanotechnologies can improve capacity and safety of lithium-ion batteries decisively, as for example through new ceramic, heat-resistant and still flexible separators and high-performance electrode materials. The company Evonik pushes the commercialization of such systems for the application in hybrid and electric vehicles as well as for stationary energy storage.

In the long run, even hydrogen seems to be a promising energy store for environmentally-friendly energy supply. Apart from necessary nanostructure adjustments, the efficient storage of hydrogen is regarded as one of the critical factors of success on the way to a possible hydrogen management. Current materials for chemical hydrogen storage do not meet the demands of the automotive industry, which requires a hydrogen-storage capacity of up to ten weight percent.

Various nanomaterials, inter alia based on nanoporous metal-organic compounds, provide development potentials, which seem to be economically realisable at least with regard to the operation of fuel cells in portable electronic devices.

Another important field is thermal energy storage. The energy demand in buildings, for example, may be significantly reduced by using phase change materials such as latent heat stores. Interesting, from an economic point of view, are also adsorption stores based on nanoporous materials like zeolites, which could be applied as heat stores in district heating grids or in industry. The adsorption of water in zeolite allows the reversible storage and release of heat.

Energy Usage

To achieve sustainable energy supply, and parallel to the optimized development of available energy sources, it is necessary to improve the efficiency of energy use and to avoid unnecessary energy consumption. This applies to all branches of industry and private households. Nanotechnologies provide a multitude of approaches to energy saving. Examples are the reduction of fuel consumption in automobiles through lightweight construction materials on the basis of nanocomposites, the optimization in fuel combustion through wear-resistant, lighter engine components and nanoparticular fuel additives or even nanoparticles for optimized tires with low rolling resistance. Considerable energy savings are realisable through tribological layers for mechanical components in plants and machines. Building technology also provides great potentials for energy savings, which could be tapped, for example, by nanoporous thermal insulation material suitably applicable in the energetic rehabilitation of old buildings. In general, the control of light and heat flux by nanotechnological components, as for example switchable glasses, is a promising approach to reducing energy consumption in buildings.

Nanogenerator

Nanogenerator is a technology that converts mechanical/thermal energy as produced by small-scale physical change into electricity. Nanogenerator has three typical approaches: piezoelectric, triboelectric, and pyroelectric nanogenerators. Both the piezoelectric and triboelectric nanogenerators can convert the mechanical energy into electricity. However, the pyroelectric nanogenerators can be used to harvest thermal energy from a time-dependent temperature fluctuation.

Piezoelectric Nanogenerator

A piezoelectric nanogenerator is an energy harvesting device converting the external kinetic energy into an electrical energy based on the energy conversion by nano-structured piezoelectric material. Although its definition may include any types of energy harvesting devices with nano-structure converting the various types of the ambient energy (e.g. solar power and thermal energy), it is used in most of times to specifically indicate the kinetic energy harvesting devices utilising nano-scaled piezoelcctric material after its first introduction in 2006.

Although still in the early stage of the development, it has been regarded as a potential breakthrough toward the further miniaturization of the conventional energy harvester, possibly leading the facile integration with the other types of energy harvester converting the different types of energy and the independent operation of mobile electronic devices with the reduced concerns for the energy source, consequently.

Mechanism

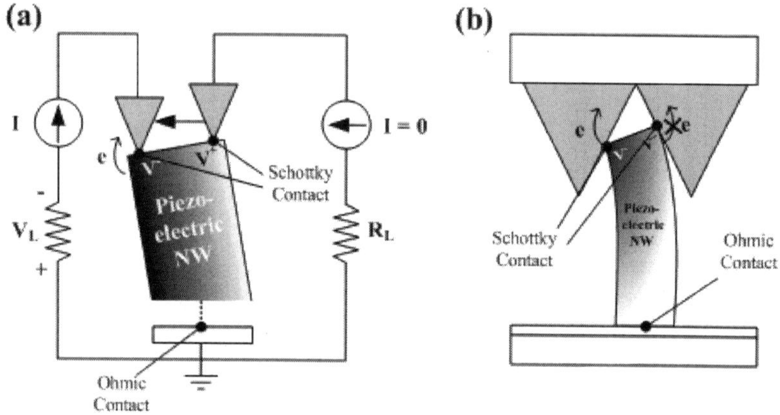

***Figure:** Working principle of nanogenerator where an individual nanowire is subjected to the force exerted perpendicular to the growing direction of nanowire. (a) An AFT tip is swept through the tip of the nanowire. Only negatively charged portion will allow the current to flow through the interface. (b) The nanowire is integrated with the counter electrode with AFT tip-like grating. As of (a), the electrons are transported from the compressed portion of nanowire to the counter electrode because of Schottky contact.*

The working principle of nanogenerator will be explained for 2 different cases: the force exerted perpendicular and parallel to the axis of the nanowire.

The working principle for the first case is explained by a vertically grown nanowire subjected to the laterally moving tip. When a piezoelectric structure is subjected to the external force by the moving tip, the deformation occurs throughout the structure. The piezoelectric effect will create the electrical field inside the nanostructure; the stretched part with the positive strain will exhibit the positive electrical potential, whereas the compressed part with the negative strain will show the negative electrical potential. This is due to the relative displacement of cations with respect to anions in its crystalline structure. As a result, the tip of the nanowire will have an electrical potential distribution on its surface, while the bottom of the nanowire is neutralized since it is grounded. The maximum voltage generated in the nanowire can be calculated by the following equation:

$$V_{\max} = \pm \frac{3}{4(\kappa_0 + \kappa)} [e_{33} - 2(1+\nu)e_{15} - 2\nu e_{31}] \frac{a^3}{l^3} \nu_{\max}$$

where κ_0 is the permittivity in vacuum, ê is the dielectric constant, e_{33}, e_{15} and e_{31} are the piezoelectric coefficients, ν is the Poisson ratio, a is the radius of the nanowire, l is the length of the nanowire and v_{max} is the maximum deflection of the nanowire's tip.

The electrical contact plays an important role to pump out charges in the surface of the tip. The schottky contact must be formed between the counter electrode and the tip of the nanowire since the ohmic contact will neutralize the electrical field generated at the tip. In order to form an effective schottky contact, the electron affinity(E_a) must be smaller than the work function(φ) of the metal composing the counter electrode. For the case of ZnO nanowire with the electron affinity of 4.5 eV, Pt (φ=6.1eV) is a suitable metal to construct the schottky contact. By constructing the schottky contact, the electrons will pass to the counter electrode from the surface of the tip when the counter electrode is in contact with the regions of the negative potential, whereas no current will be generated when it is in contact with the regions of the positive potential, in the case of n-type semiconductive nanostructure (p-type semiconductive structure will exhibit the reversed phenomenon since the hole is mobile in this case). The formation of the schottky contact also contributes to the generation of direct current output signal consequently.

For the second case, a model with a vertically grown nanowire stacked between the ohmic contact at its bottom and the schottky contact at its top is considered. When the force is applied toward the tip of the nanowire, the uniaxial compressive is generated in the nanowire. Due to the piezoelectric effect, the tip of the nanowire will have a negative piezoelectric potential, increasing the Fermi level at the tip. Since the electrons will then flow from the tip to the bottom through the external circuit as a result, the positive electrical potential will be generated at the tip. The schottky contact will barricade the electrons being transported through the interface, therefore maintaining the potential at the tip. As the force is removed, the piezoelectric effect diminishes, and the electrons will be flowing back to the top in order to neutralize the positive potential at the tip. The second case will generate alternating current output signal.

Geometrical Configuration

Depending on the configuration of piezoelectric nanostructure, the most of the nanogenerator can be categorized into 3 types: VING, LING and "NEG". Still, there is a configuration that do not fall into the aforementioned categories, as stated in other type.

Vertical Nanowire Integrated Nanogenerator (VING).

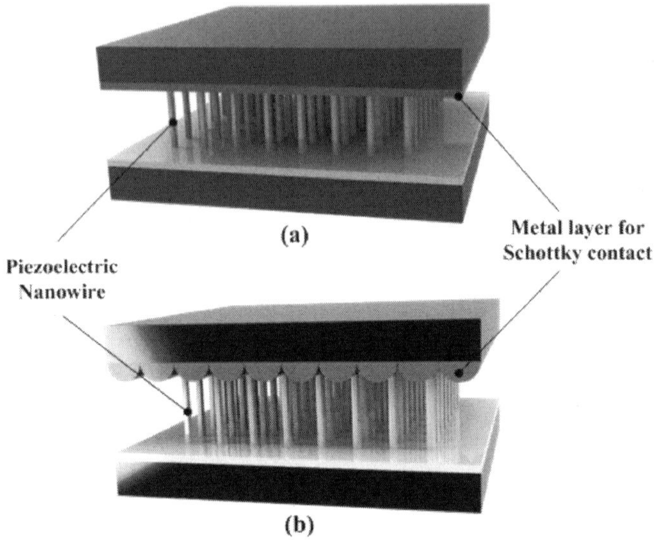

Figure: *Schematic view of typical Vertical nanowire Integrated Nanogenerator, (a) with full contact, and (b) with partial contact. Note that the grating on the counter electrode is important in the latter case.*

VING is a 3-dimensional configuration consisting of a stack of 3 layers in general, which are the base electrode, the vertically grown piezoelectric nanostructure and the counter electrode. The piezoelectric nanostructure is usually grown from the base electrode by various synthesizing techniques, which are then integrated with the counter electrode in full or partial mechanical contact with its tip.

After Professor Zhong Lin Wang of the Georgia Institute of Technology has introduced a basic configuration of VING in 2006 where he used a tip of atomic force microscope (AFM) to induce the deformation of a single vertical ZnO nanowire, the first development of VING is followed in 2007. The first VING utilises the counter electrode with the periodic surface grating resembling the arrays of AFM tip as a moving electrode. Since the counter electrode is not in full contact with the tips of the piezoelectric nanowire, its motion in-plane or out-of-plane occurred by the external vibration induces the deformation of the piezoelectric nanostructure, leading to the generation of the electrical potential distribution inside each individual nanowire. It should be noted that the counter electrode is coated with the metal forming the schottky contact with the tip of the nanowire, where only the compressed portion of piezoelectric nanowire would allow the accumulated electrons pass through the barrier between its tip and the counter electrode, in case of n-type nanowire. The switch-on and –off characteristic of this configuration shows its capability of generating direct current generation without any requirement for the external rectifier.

In VING with partial contact, the geometry of the counter electrode plays an important role. The flat counter electrode would not induce the sufficient deformation of the piezoelectric nanostructures, especially when the counter electrode moves by in-plane mode. After the basic geometry resembling the array of AFM tips, a few other approaches have been followed for facile development of the counter electrode. Professor Zhong Lin Wang's group have generated counter electrode composed of ZnO nanorods utilising the similar technique used for synthesizing ZnO nanowire array. Professor Sang-Woo Kim's group of Sungkyunkwan University (SKKU) and Dr. Jae-Young Choi's group of Samsung Advanced Institute of Technology (SAIT) in South Korea introduced bowl-shaped transparent counter electrode by combining anodized aluminum and the electroplating technology. They also have developed the other type of the counter electrode by using networked single-walled carbon-nanotube (SWNT) on the flexible substrate, which is not only effective for energy conversion but also transparent.

The other type of VING has been also suggested. While it shares the identical geometric configuration with the aforementioned, such a VING has full mechanical contact between the tips of the nanowires and the counter electrode. This configuration is effective for application where the force is exerted in the vertical direction (toward the c axis of the piezoelectric nanowire), and it generates alternating current (AC) unlike VINGs with partial contact.

Lateral Nanowire Integrated Nanogenerator (LING).

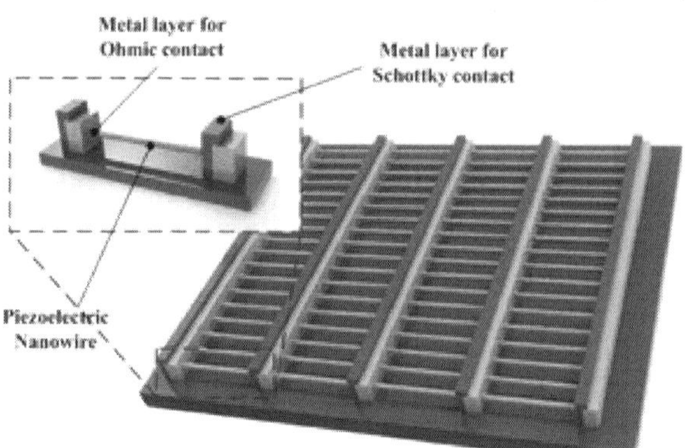

Figure: *Schematic view of typical Lateral nanowire Integrated Nanogenerator*

LING is a 2-dimensional configuration consisting of three parts: the base electrode, the laterally grown piezoelectric nanostructure and the metal electrode for schottky contact. In most of cases, the thickness of the substrate film is much thicker than the diameter of the piezoelectric nanostructure, so the individual nanostructure is subjected to the pure tensile strain.

LING is an expansion of single wire generator (SWG), where a laterally aligned nanowire is integrated on the flexible substrate. SWG is rather a scientific configuration used for verifying the capability of electrical energy generation of a piezoelectric material and is widely adopted in the early stage of the development.

As of VINGs with full mechanical contact, LING generates AC electrical signal. The output voltage can be amplified by constructing an array of LING connected in series on the single substrate, leading the constructive addition of the output voltage. Such a configuration may lead to the practical application of LING for scavenging large-scale power, for example, wind or ocean waves.

Nanocomposite Electrical Generators (NEG).

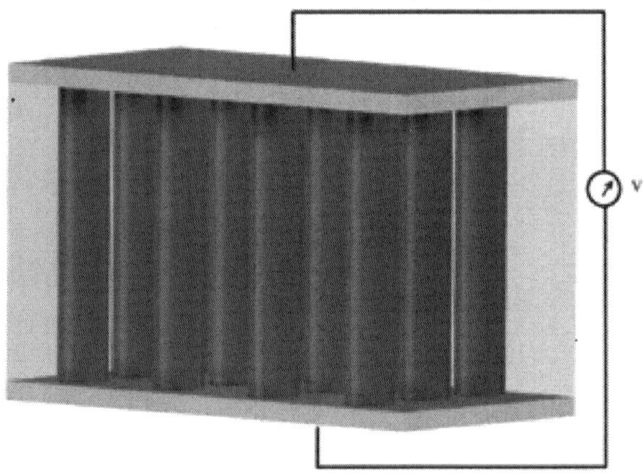

Figure: *Schematic view of typical Nanocomposite Electrical Generator*

"NEG" is a 3-dimensional configuration consisting three main parts: the metal plate electrodes, the vertically grown piezoelectric nanostructure and the polymer matrix which fills in between in the piezoelectric nanostructure.

NEG was introduced by Momeni et al. It was shown that NEG has a higher efficiency compared to original nanogenerator configuration which a ZnO nanowire will be bended by an AFM tip. It is also shown that it provides an energy source with higher sustainability.

Other type. The fabric-like geometrical configuration has been suggested by Professor Zhong Lin Wang in 2008. The piezoelectric nanowire is grown vertically on the two microfibres in its radial direction, and they are twined to form a nanogenerator. One of the microfibres is coated with the metal to form a schottky contact, serving as the counter electrode of VINGs. As the movable microfibre is stretched, the deformation of the nanostructure occurs on the stationary microfibre, resulting in the voltage generation. Its working principle is identical to VINGs with partial mechanical contact, thus generating DC electrical signal.

Materials

Among various piezoelectric materials studied for the nanogenerator, many of the researches have been focused on the materials with wurtzite structure such as ZnO, CdS and GaN. The greatest advantage of theses material arises from the facile and cost-effective fabrication technique, hydrothermal synthesis. Since the hydrothermal synthesis can be conducted in a low temperature environment under 100°C in addition to vertical and crystalline growth, these materials can be integrated in various substrates with reduced concern for its physical characteristics such as a melting temperature.

Endeavours for enhancing the piezoelectricity of the individual nanowire also led to the development of other piezoelectric materials based on Wurtzite structure. Professor Zhong Lin Wang of Georgia Institute of Technology introduced p-type ZnO nanowire. Unlike the n-type semiconductive nanostructure, the mobile particle in p-type is a hole, thus the schottky behaviour is reversed from that of n-type case; the electrical signal is generated from the portion of the nanostructure where the holes are accumulated. It is experimentally proved that p-type ZnO nanowire can generate the output signal near 10 times that of n-type ZnO nanowire.

From the idea that the material with perovskite structure is known to have more effective piezoelectric characteristic compared to that with wurtzite structure, Barium titanate ($BaTiO_3$) nanowire has been also studied by Professor Min-Feng Yu of University of Illinois at Urbana Champaign. The output signal is found to be more than 16 time that from a similar ZnO nanowire.

Professor Liwei Lin of University of California at Berkeley has suggested that PVDF can be also applied to form a nanogenerator. Being a polymer, PVDF utilises a near-field electrospinning for its fabrication, which is rather a different technique compared to other materials. The nanofibre can be directly written on the substrate controlling the process, and this technique is expected to be applied for forming self-powered textile based on nanofibre.

Comparison of the reported materials by 2010 is given in the following table.

Material	*Type*	*Geometry*	*Output voltage*	*Output power*	*Synthesis*	*Researched at*
ZnO (n-type)	Wurtzite	D: ~100 nm, L: 200~500 nm	V_P=~9 mV @ R=500 MΩ	~0.5 pW per cycle (estimated)	CVD, hydrothermal process	Georgia Tech.
ZnO (p-type)	Wurtzite	D: ~50 nm, L: ~600 nm	V_P=50~90 mV @ R=500 MΩ	5~16.2 pW per cycle (calculated)	CVD	Georgia Tech.
ZnO-ZnS	Wurtzite (Heterostructure)	Not stated	V_P=~6 mV @ R=500 MΩ	~0.1 pW per cycle (calculated)	Thermal evaporation and etching	Georgia Tech.
GaN	Wurtzite	D: 25~70 nm, L: 10~20 μm	V_{avg}=~20 mV, V_{max}=~0.35 V@ R=500 MΩ	~0.8 pW per cycle (average, calculated)	CVD	Georgia Tech.
CdS	Wurtzite	D: ~100 nm, L: 1 μm	V_P=~3 mV	Not stated	PVD, Hydrothermal Process	Georgia Tech.
$BaTiO_3$	Pervoskite	D: ~280 nm, L: ~15 μm	V_P=~25 mV @ R=100 MΩ	~0.3 aJ per cycle (stated)	High temperature chemical reaction	UIUC
PVDF	Polymer	D: 0.5~6.5 μm, L: 0.1~0.6 mm	V_P=5~30 mV	2.5 pW~90 pW per cycle (calculated)	Electro spinning	UC Berkeley

Applications

Nanogenerator is expected to be applied for various applications where the periodic kinetic energy exists, such as wind and ocean waves in a large scale to the muscle movement by the beat of a heart or inhalation of lung in a small scale. The further feasible applications are as follows.

Self-powered nano/micro devices. One of the feasible applications of nanogenerator is an independent or a supplementary energy source to nano/micro devices consuming relatively low amount of energy in a condition where the kinetic energy is supplied continuously. One of example has been introduced by Professor Zhong Lin Wang's group in 2010 by the self-powered pH or UV sensor integrated VING with an output voltage of 20~40 mV onto the sensor.

Still, the converted electrical energy is relatively small for operating nano/micro devices; therefore the range of its application is still bounded as a supplementary energy source to the battery. The breakthrough is being sought by combining the nanogenerator with the other types of energy harvesting devices, such as solar cell or biochemical energy harvester. This approach is expected to contribute to the development of the energy source suitable for the application where the independent operation is crucial, such as Smartdust.

Smart Wearable Systems. The outfit integrated or made of the textiles with the piezoelectric fibre is one of the feasible applications of the nanogenerator. The kinetic energy from the human body is converted to the electrical energy through the piezoelectric fibres, and it can be possibly applied to supply the portable electronic devices such as health-monitoring system attached with the Smart Wearable Systems. The nanogenerator such as VING can be also easily integrated in the shoe employing the walking motion of human body.

Another similar application is a power-generating artificial skin. Professor Zhong Lin Wang's group has shown the possibility by generating AC voltage of up to 100 mV from the flexible SWG attached to the running hamster.

Transparent and Flexible Devices. Some of the piezoelectric nanostructure can be formed in various kinds of substrates, such as flexible and transparent organic substrate. The research groups in SKKU (Professor Sang-Woo Kim's group) and SAIT (Dr. Jae-Young Choi's group) have developed the transparent and flexible nanogenerator which can be possibly used for self-powered tactile sensor and anticipated that the development may be extended to the energy-efficient touch screen devices. Their research focus is being extended to enhance the transparency of the device and the cost-effectiveness by substituting Indium-Tin-Oxide (ITO) electrode with a graphene layer.

Implantable Telemetric Energy Receiver. The nanogenerator based on ZnO nanowire can be applied for implantable devices since ZnO not only is bio-compatible but also can be synthesized upon the organic substrate, rendering the nanogenerator bio-compatible in overall. The implantable device integrated with the nanogenerator can be operated by receiving the external ultrasonic vibration outside the human body, which is converted to the electrical energy by the piezoelectric nanostructure.

Triboelectric Nanogenerator

A triboelectric nanogenerator is an energy harvesting device that converts the external the ambient mechanical energy into electricity by a conjunction of triboelectric effect and electrostatic induction. This new type of nanogenerator was firstly demonstrated in Prof. Zhong Lin Wang's group at Georgia Institute of Technology in the year of 2012. As for this power generation unit, in the inner circuit, a potential is created by the triboelectric effect due to the charge transfer between two thin organic/inorganic films that exhibit opposite tribo-polarity; in the outer circuit, electrons are driven to flow between two electrodes attached on the back sides of the films in order to balance the potential. Since the most useful materials for TENG are organic, it is also named organic nanogenerator, which is the first of using organic materials for harvesting mechanical energy.

Ever since the first report of the TENG in January 2012, the output power density of TENG has been improved for five orders of magnitude within 12 months. The area power density reaches 313 W/m2, volume density reaches 490 kW/m3, and a conversion efficiency of ~60% has been demonstrated. Besides the unprecedented output performance, this new energy technology also has a number of other advantages, such as low cost in manufacturing and fabrication, excellent robustness and reliability, environmental-friendly, and so on. The triboelectric nanogenerator can be applied to harvest all kind mechanical energy that is available but wasted in our daily life, such as human motion, walking, vibration, mechanical triggering, rotating tire, wind, flowing water and more.

The triboelectric nanogenerator has three basic operation modes: vertical contact-separation mode, in-plane sliding mode, and single-electrode mode. They has different characteristics and are suitable for different applications

Basic Modes and Mechanisms

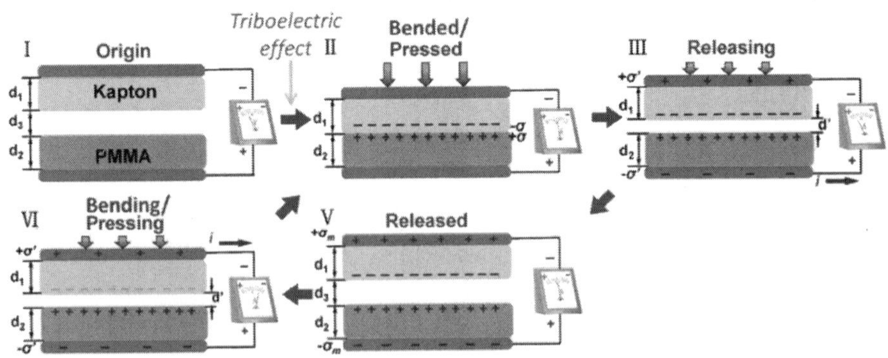

Figure: *Vertical contact-separation mode of triboelectric nanogenerator*

Vertical Contact-Separation Mode: The working mechanism of the triboelectric nanogenerator can be described as the periodic change of the potential difference induced by the cycled separation and re-contact of the opposite triboelectric charges on the inner surfaces of the two sheets. When a mechanical agitation is applied onto the device to bend or press it, the inners surfaces of the two sheets will get into close contact and the charge transfer will begin, leaving one side of the surface with positive charges and the other with negative charges. This is just the triboelectric effect. When the deformation is released, the two surfaces with opposite charges will separate automatically, so that these opposite triboelectrc charges will generate an electric field in between and thus induce a potential difference across the top and bottom electrodes. In order to screen this potential difference, the electrons will be driven to flow from one electrode to the other through the external load. The electricity generated in this process will continue until the potentials of the two electrodes get back to even again. Subsequently, when the two sheets are pressed towards each other again, the triboelectric-charge-induced potential difference will begin to decrease to zero, so that the transferred charges will flow back through the external load, to generate another current pulse in the opposite direction. When this periodic mechanical deformation lasts, the alternating current (AC) signals will be continuously generated.

As for the pair of materials getting in contact and generating triboelectric charges, at least one of them need to be an insulator, so that the triboelectric charges cannot be conducted away but will remain on the inner surface of the sheet. Then, these immobile triboelectric charges can induce AC electricity flow in the external load under the periodic distance change.

Lateral Sliding Mode

There are two basic friction processes: normal contact, and lateral sliding. We demonstrated here a TENG that is designed based on the in-plane sliding between the two surfaces in lateral direction. With an intensive triboelectrification facilitated by sliding friction, a periodic change in the contact area between two surfaces leads to a lateral separation of the charge centres, which creates a voltage drop for driving the flow of electrons in the external load. The sliding-induced electricity generation mechanism is schematically depicted in the figure. In the original position, the two polymeric surfaces fully overlap and intimately contact with each other. Because of the large difference in the ability to attract electrons, the triboelectrification will leave one surface with net positive charges and the other with net negative charges with equal density. Since the tribo-charges on the insulators will only distribute in the surface layer and will not be leaked out for an extended period of time, the separation between the positively-charged surface and negatively-charged surface is negligible at this overlapping position, and thus there will be little electric potential drop across the two electrodes. Once the top plate with the positively-charged surface starts to slide outward, the in-plane charge separation is initiated due to the decrease in contact surface area.

The separated charges will generate an electric field pointing from the right to the left almost parallel to the plates, inducing a higher potential at the top electrode. This potential difference will drive a current flow from the top electrode to the bottom electrode in order to generate an electric potential drop that cancels the tribo-charge-induced potential. Because the vertical distance between the electrode layer and the tribo-charged polymeric surface is negligible compared to the lateral charge separation distance, the amount of the transferred charges on the electrodes approximately equals to the amount of the separated charges at any sliding displacement.

Thus, the current flow will continue with the continuation of the on-going sliding process that keeps increasing the separated charges, until the top plate fully slides out of the bottom plate and the tribo-charged surfaces are entirely separated. The measured current should be determined by the rate at which the two plates are being slid apart. Subsequently, when the top plate is reverted to slide backwards, the separated charges begins to get in contact again but no annihilation due to the insulator nature of the polymer materials.

The redundant transferred charges on the electrodes will flow back through the external load with the increase of the contact area, in order to keep the electrostatic equilibrium. This will contribute to a current flow from the bottom electrode to the top electrode, along with the second half cycle of sliding. Once the two plates reach the overlapping position, the charged surfaces get into fully contact again. There will be no transferred charges left on the electrode, and the device returns to the first state. In this entire cycle, the processes of sliding outwards and inwards are symmetric, so a pair of symmetric alternating current peaks should be expected.

The mechanism of in-plane charge separation can work in either one directional sliding between two plates or in rotation mode. In the sliding mode, introducing linear grating or circular segmentation on the sliding surfaces is an extremely efficient means for energy harvesting. With such structures, two patterned triboelectric surfaces can get to fully mismatching position through a displacement of only a grating unit length rather than the entire length of the TENG so that it dramatically increase the transport efficiency of the induced charges.

Single-Electrode Mode

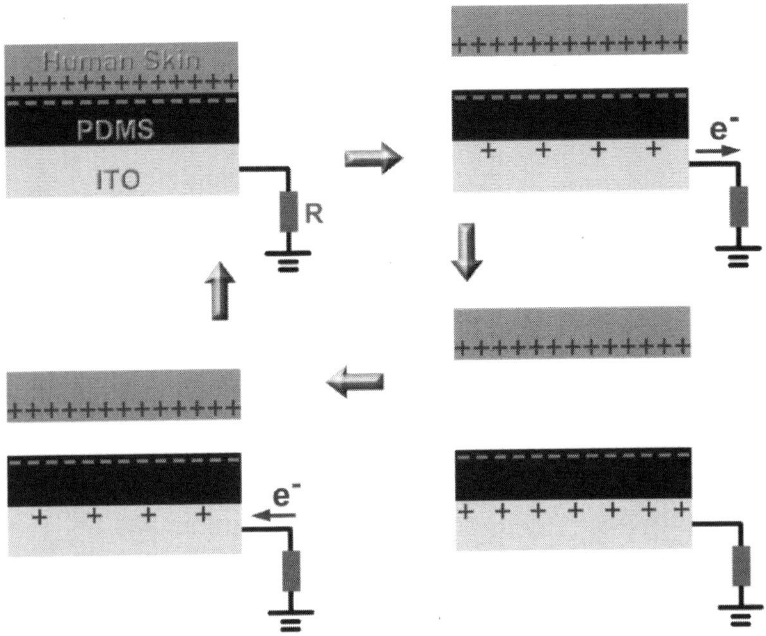

***Figure:** Single-electrode mode of triboelectric nanogenerator*

A single-electrode based triboelectric nanogenerator is introduced as a more practical and feasible design for some applications such as finger-tip driven triboelectric nanoagenerator. The working principle of the single-electrode TENG is schematically shown in the figure by the coupling of contact electrification and electrostatic induction. In the original position, the surfaces of skin and PDMS fully contact with each other, resulting in charge transfer between them. According to the triboelectric series, electrons were injected from the skin to the PDMS since the PDMS is more triboelectrically negative than skin, which is the contact electrification process. The produced triboelectric charges with opposite polarities are fully balanced/screened, leading to no electron flow in the external circuit. Once a relative separation between PDMS and skin occurs, these triboelectric charges cannot be compensated. The negative charges on the surface of the PDMS can induce positive charges on the ITO electrode, driving free electrons to flow from the ITO electrode to ground. This electrostatic induction process can give an output voltage/current signal if the distance separating between the touching skin and the bottom PDMS is appreciably comparable to the size of the PDMS film. When negative triboelectric charges on the PDMS are fully screened from the induced positive charges on the ITO electrode by increasing the separation distance between the PDMS and skin, no output signals can be observed, as illustrated. Moreover, when the skin was reverted to approach the PDMS, the induced positive charges on the ITO electrode decrease and the electrons will flow from ground to the ITO electrode until the skin and PDMS fully contact with each other again, resulting in a reversed output voltage/current signal. This is a full cycle of electricity generation process for the TENG in contact-separation mode.

Applications

TENG is a physical process of converting mechanical agitation to an electric signal through the triboelectrification (in inner circuit) and electrostatic induction processes (in outer circuit). This basic process has been demonstrated for two major applications. The first application is energy harvesting with a particular advantage of harvesting mechanical energy. The other application is to serve as a self-powered active sensor, because it does not need an external power source to drive.

Harvesting Vibration Energy

Vibration is one of the most popular phenomena in our daily life, from walking, voices, engine vibration, automobile, train, aircraft, wind and many more. It exists almost everywhere and at all the time. Harvesting vibration energy is of great value especially for powering mobile electronics. The following Based on the fundamental principles of triboelectric nanogenerators, various technologies have been demonstrated for harvesting vibration energy. This application of triboelectric nanogenerator has been demonstrated in the following aspects: 1. Cantilever based technique is a classical approach for harvesting mechanical energy, especially for MEMS. By designing the contact surface of a cantilever with the top and bottom surfaces during vibration, TENG has been demonstrated for harvesting ambient vibration energy based on the contact-separation mode. 2. To harvest the energy from a backpack, we demonstrated a rationally designed TENG with integrated rhombic gridding, which greatly improved the total current output owing to the structurally multiplied unit cells connected in parallel. 3. With the use of 4 supporing springs, a harmonic resonator based TENG has been fabricated based on the resonance induced contact-separation between the two triboelectric materials, which has been used to harvest vibration energy from an automobile engine, a sofa and a desk. 4. Recently, a three-dimensional triboelectric nanogenerator (3D-TENG) has been designed based on a hybridization mode of conjunction the vertical contact-separation mode and the in-plane sliding mode.36 The innovative design facilitates harvesting random vibration energy in multiple directions over a wide bandwidth. The 3-D TENG is designed for harvesting ambient vibration energy, especially at low frequencies, under a range of conditions in daily life, thus, opening the applications of TENG in environmental/infrastructure monitoring, charging portable electronics and internet of things.

Harvesting Energy from Human Body Motion

Since there is abundant mechanical energy generated on human bodies in people's everyday life, we can make use of the triboelectric nanogenerator to convert this amount of mechanical energy into electricity, for charging portable electronics and biomedical applications. This will help to greatly improve the convenience of people's life and expand the application of the personal electronics. A packaged power-generating insole with built-in flexible multi-layered triboelectric nanogenerators has been demonstrated, which enable harvesting mechanical pressure during normal walking. The TENG used here relies on the contact-separation mode and is effective in responding to the periodic compression of the insole. Using the insole as a direct power source, we develop a fully packaged self-lighting shoe that has broad applications for display and entertainment purposes. A TENG can be attached to the inner layer of a shirt for harvesting energy from body motion. Under the generally walking, the maximum output of voltage and current density are up to 17 V and 0.02 $\mu A/cm^2$, respectively. The TENG with a single layer size of 2 cm×7 cm×0.08 cm sticking on the clothes was demonstrated as a sustainable power source that not only can directly light up 30 light-emitting diodes (LEDs), but also can charge a lithium ion battery by persistently clapping clothes.

Self-Powered Active Strain/Force Sensors

An triboelectric nanogenerator automatically generates an output voltage and current once it is mechanically triggered. The magnitude or the output signal signifies the impact of the mechanical deformation and its time-dependent behaviour. This is the basic principle of the TENG can be applied as a self-powered pressure sensor. The voltage-output signal can reflect the applied pressure induced by a droplet of water. All types of TENGs have a high sensitivity and fast response to the external force and show as a sharp peak signal. Furthermore, the response to the impact of a piece of feather (20 mg, ~0.4 Pa in contact pressure) can be detected. The sensor signal can delicately show these details of the entire process. The existing results show that our sensor can be applied for measuring the subtle pressure in real life.

In a case that we make a matric array of the triboelectric nanogenerators, a large-area, and self-powered pressure map applied on a surface can be realised. The response of the TENG array with local pressure was measured through a multi-channel measurement system. There are two types of output signals from the TENG: open circuit voltage and short circuit current. The Open circuit voltage is only dictated by the final configuration of the TENG after applying a mechanical triggering, so that it is a measure of the magnitude of the deformation, which is attributed to the static information to be provided by TENG. The output current depends on the rate at which the induced charge would flow, so that the current signal is more sensitive to the dynamic process of how the mechanical triggering is applied.

The active pressure sensor and the integrated sensor array based on the triboelectric effect have several advantages over conventional passive pressure sensors. First, the active sensor is capable of both static pressure sensing using the open-circuit voltage and dynamic pressure sensing using the short-circuit current, while conventional sensors are usually incapable of dynamic sensing to provide the loading rate information. Second, the prompt response of both static and dynamic sensing enables the revealing of details about the loading pressure. Third, the detection limit of the TENG for dynamic sensing is as low as 2.1 Pa, owing to the high output of the TENG. Fourth, the active sensor array presented in this work has no power consumption and could even be combined with its energy harvesting functionality for self-powered pressure mapping. Future works in this field involve the miniaturization of the pixel size to achieve higher spatial resolution, and the integration of the TEAS matrix onto fully flexible substrate for shape-adaptive pressure imaging.

Self-Powered Active Chemical Sensors

As for triboelectric nanogenerators, maximizing the charge generation on opposite sides can be achieved by selecting the materials with the largest difference in the ability to attract electrons and changing the surface morphology. In such a case, the output of the TENG depends on the type and concentration of molecules adsorbed on the surface of the triboelectric materials, which can be used for fabricating chemical and biochemical sensors. As an example, the performance of the TENG depends on the assembly of Au nanoparticles (NPs) onto the metal plate. These assembled Au NPs not only act as steady gaps between the two plates at strain free condition, but also enable the function of enlarging the contact area of the two plates, which will increase the electrical output of the TENG. Through further modification of 3-

mercaptopropionic acid (3-MPA) molecules on the assembled Au NPs, the high-output nanogenerator can become a highly sensitive and selective nanosensor toward Hg^{2+} ions detection because of the different triboelectric polarity of Au NPs and Hg^{2+} ions. With its high sensitivity, selectivity and simplicity, the TENG holds great potential for the determination of Hg^{2+} ions in environmental samples. The TENG is a future sensing system for unreachable and access-denied extreme environments. As different ions, molecules, and materials have their unique triboelectric polarities, we expect that the TENG can become either an electrical turn-on or turn-off sensor when the analytes are selectively binding to the modified electrode surface. We believe this work will serve as the stepping stone for related TENG studies and inspire the development of TENG toward other metal ions and biomolecules such as DNA and proteins in the near future.

Choice of Mterials and Surface Structures

Almost any materials we know have triboelectrification effect, from metal, to polymer, to silk and to wood, almost everything. All of these materials can be candidates for fabricating TENGs, so that the materials choices for TENG are huge. However, the ability of a material for gaining/losing electron depends on its polarity. John Carl Wilcke published the first triboelectric series in a 1757 on static charges. A material towards the bottom of the series, when touched to a material near the top of the series, will attain a more negative charge. The further away two materials are from each other on the series, the greater the charge transferred. Beside the choice of the materials in the triboelectric series, the morphologies of the surfaces can be modified by physical techniques with the creation of pyramids-, square- or hemisphere-based micro- or nano-patterns, which are effective for enhancing the contact area and possibly the triboelectrification. However, the created bumpy structure on the surface may increase the friction force, which may possibly reduce the energy conversion efficiency of the TENG. Therefore, an optimization has to be designed for maximizing the conversion efficiency.

The surfaces of the materials can be functionalized chemically using various molecules, nanotubes, nanowires or nanoparticles, in order to enhance the triboelectrification effect. Surface functionalization can largely change the surface potential. The introduction of nanostructures on the surfaces can change the local contact characteristics, which may improve the triboelectrification. This will involve a large amount of studies for testing a range of materials and a range of available nanostructures.

Besides these pure materials, the contact materials can be made of composites, such embedding nanoparticles in polymer matrix. This not only changes the surface electrification, but also the permittivity of the materials so that they can be effective for electrostatic induction. Therefore, there are numerous ways for enhancing the performance of the TENG from the materials point of view. This gives an excellent opportunity for chemists and materials scientists to do extensive study both in the basic science and in practical application. In contrast, materials systems for solar cell and thermal electric, for example, are rather limited, and there are not very many choices for high performance devices.

The surfaces of the materials can be functionalized chemically using various molecules, nanotubes, nanowires or nanoparticles, in order to enhance the triboelectrification effect. Surface functionalization can largely change the surface potential. The introduction of nanostructures on the surfaces can change the local contact characteristics, which may improve the triboelectrification. This will involve a large amount of studies for testing a range of materials and a range of available nanostructures.

Besides these pure materials, the contact materials can be made of composites, such embedding nanoparticles in polymer matrix. This not only changes the surface electrification, but also the permittivity of the materials so that they can be effective for electrostatic induction.

Therefore, there are numerous ways for enhancing the performance of the TENG from the materials point of view. This gives an excellent opportunity for chemists and materials scientists to do extensive study both in the basic science and in practical application. In contrast, materials systems for solar cell and thermal electric, for example, are rather limited, and there are not very many choices for high performance devices.

Pyroelectric Nanogenerator

A pyroelectric nanogenerator is an energy harvesting device converting the external thermal energy into an electrical energy by using nano-structured pyroelectric materials. Usually, harvesting thermoelectric energy mainly relies on the Seebeck effect that utilises a temperature difference between two ends of the device for driving the diffusion of charge carriers. However, in an environment that the temperature is spatially uniform without a gradient, such as in outdoor in our daily life, the Seebeck effect can not be used to harvest thermal energy from a time-dependent temperature fluctuation. In this case, the pyroelectric effect has to be the choice, which is about the spontaneous polarization in certain anisotropic solids as a result of temperature fluctuation. The first pyroelectric nanogenerator was introduced by Prof. Zhong Lin Wang at Georgia Institute of Technology in 2012. By harvesting the waste heat energy, this new type of nanogenerator has the potential applications such as wireless sensors, temperature imaging, medical diagnostics, and personal electronics.

Mechanism

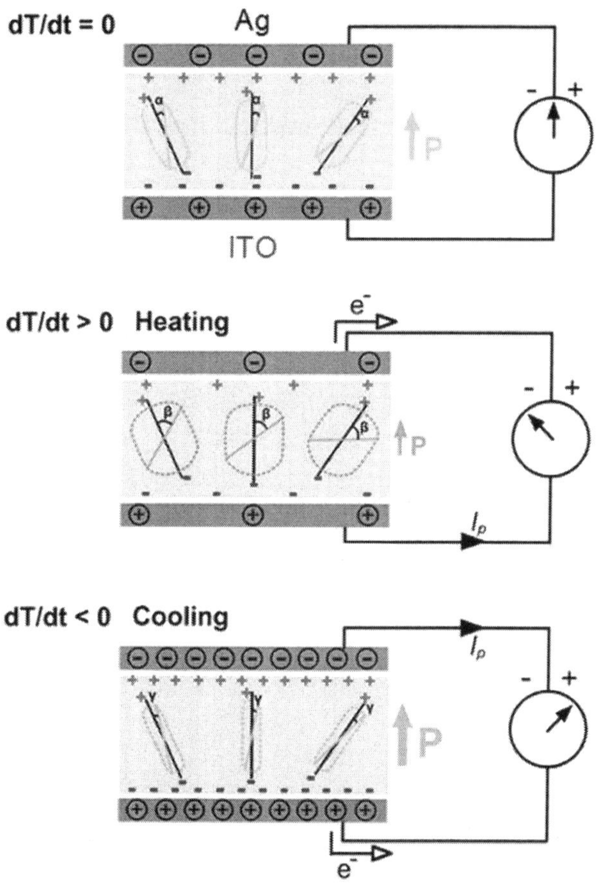

Figure: The mechanism of the pyroelectric nanogenerator based on a composite structure of pyroelectric nanowries.(a-c) Schematic diagrams of the pyroelectric nanogenerator with negative electric dipoles under room temperature (a), heated (b) and cooled (c) conditions. The angles marked in the diagrams represent the degrees to which the dipole would oscillate as driven by statistical thermal fluctuations.

The working principle of pyroelectric nanogenerator will be explained for 2 different cases: the primary pyroelectric effect and the secondary pyroelectric effect. The working principle for the first case is explained by the primary pyroelectric effect, which describes the charge produced in a strain-free case. The primary pyroelectric effect dominates the pyroelectric response in PZT, BTO, and some other ferroelectric materials. The mechanism is based on the thermally induced random wobbling of the electric dipole around its equilibrium axis, the magnitude of which increases with increasing temperature. Due to thermal fluctuations under room temperature, the electric dipoles will randomly oscillate within a degree from their respective aligning axes. Under a fixed temperature, the total average strength of the spontaneous polarization form the electric dipoles is constant, resulting in no output of the pyroelectric nanogenerator. If we apply a change in temperature in the nanogenerator from room temperature to a higher temperature, the increase in temperature will result in that the electric dipoles oscillate within a larger degree of spread around their respective aligning axes. The total average spontaneous polarization is decreased due to the spread of the oscillation angles. The quantity of induced charges in the electrodes are thus reduced, resulting in a flow of electrons. If the nanogenerator is cooled instead of heated, the spontaneous polarization will be enhanced since the electric dipoles oscillate within a smaller degree of spread angles due to the lower thermal activity. The total magnitude of the polarization is increased and the amount of induced charges in the electrodes are increased. The electrons will then flow in an opposite direction.

For the second case, the obtained pyroelectric response is explained by the secondary pyroelectric effect, which describes the charge produced by the strain induced by thermal expansion. The secondary pyroelectric effect dominates the pyroelectric response in ZnO, CdS, and some other wurzite-type materials. The thermal deformation can induce a piezoelectric potential difference across the material, which can drive the electrons to flow in the external circuit. The output of the nanogenerator is associated with the piezoelectric coefficient and the thermal deformation of the materials. The output current I of the pyroelectric nanogenerators can be determined by the equation of $I = pA(dT/dt)$, where p is the pyroelectric coefficient, A is the effective area of the NG, dT/dt is the rate of change in temperature.

Applications

Pyroelectric nanogenerator is expected to be applied for various applications where the time-dependent temperature fluctuation exists. One of the feasible applications of the pyroelectric nanogenerator is used as an active sensor, which can work without a battery. One example has been introduced by Professor Zhong Lin Wang's group in 2012 by using a pyroelectric nanogenerator as the self-powered temperature sensor for detecting a change in temperature, where the response time and reset time of the sensor are about 0.9 and 3 s, respectively. In general, the pyroelectric nanogenerator gives a high output voltage, but the output current is small. It not only can be used as a potential power source, but also as an active sensor for measuring temperature variation.

Micropower

Micropower describes the work that researchers at several universities are doing to develop very small electric generators and prime movers or devices to convert heat or motion to electricity, for use close to the generator. These devices offer the promise of a power source for portable electronic devices which is lighter weight and has a longer operating time than batteries. Micropower differs from microgeneration in being concerned with universities, mobility and milliwatts or watts rather than individuals, homes and kilowatts.

Microturbine Technology

The components of any turbine engine: the gas compressor, the combustion chamber, and the turbine rotor itself, are fabricated from etched silicon, much like integrated circuits. The technology holds the promise of ten times the operating time of a battery of the same weight as the micropower unit, and similar efficiency to large utility gas turbines. Researchers at Massachusetts Institute of Technology have thus far succeeded in fabricating the parts for such a micro turbine out of six etched and stacked silicon wafers, and are working towards combining them into a functioning engine about the size of a U.S quarter coin.

Researchers at Georgia Tech have built a micro generator 10 mm wide, which spins a magnet above an array of coils fabricated on a silicon chip. The device spins at 100,000 revolutions per minute, and produces 1.1 watts of electrical power, sufficient to operate a cell phone. Their goal is to produce 20 to 50 watts, sufficient to power a laptop computer.

Scientists at Lehigh University are developing a hydrogen generator on a silicon chip, which can convert methanol, diesel, or gasoline into fuel for a microengine or a miniature fuel cell.

Professor Sanjeev Mukerjee of Northeastern University's chemistry department is developing fuel cells for the military which will burn hydrogen to power portable electronic equipment such as night vision goggles, computers, and communication equipment. In his system, a cartridge of methanol would be used to produce hydrogen to run a small fuel cell for up to 5,000 hours. It would be lighter than rechargeable batteries needed to provide the same power output, with a longer run time. Similar technology could be improved and expanded in future years to power automobiles.

The National Academies' National Research Council recommended in a 2004 report that the U.S. Army should investigate such micropower sources for powering electronic equipment to be carried by soldiers in the future, since batteries sufficient to power the computers, sensors, and communications devices would add considerable weight to the burden of infantry soldiers.

The Future Warrior Concept of the U.S. Army envisions a 2 to 20 watt micro turbine fuelled by a liquid hydrocarbon being used to power communications and wearable heating/cooling equipment for up to six days on 10 ounces of fuel.

Other Microgenerator/Nanogenerator Technology

Professor Orest Symko of the University of Utah physics department and his students developed Thermal Acoustic Piezo Energy Conversion (TAPEC), devices of a cubic inch (16 cubic centimetres) or so which convert waste heat into acoustic resonance and then into electricity. It would be used to power microelectromechanical systems, or MEMS. The research was funded by the U.S. Army. Symko was to present a paper at the Acoustical Society of America. June 8, 2007. Researchers at MIT developed the first micro-scale piezoelectric energy harvester using thin film PZT in 2005. Arman Hajati and Sang-Gook Kim invented the Ultra Wide-Bandwidth micro-scale piezoelectric energy harvesting device by exploiting the nonlinear stiffness of a doubly clamped microelectromechanical systems (MEMSs) resonator. The stretching strain in a doubly clamped beam shows a nonlinear stiffness, which provides a passive feedback and results in amplitude-stiffened Duffing mode resonance.

Professor Zhong Lin Wang of the Georgia Institute of Technology said his team of investigators had developed a "nanometre-scale generator (which) is based on arrays of vertically aligned zinc oxide nanowires that move inside a "zigzag" plate electrode." Built into shoes, it could generate electricity from walking to power small electronic devices. It could also be powered by blood flow to power biomedical devices. Per an account of the device which appeared in the journal Science , bending of the zinc oxide nanowire arrays produces an electric field by the piezoelectric properties of the material. The semiconductor properties of the device create a Schottky barrier with rectifying capabilities. The generator is estimated to be 17% to 30% efficient in converting mechanical motion into electricity. This could be used to power biomedical devices which have wireless transmission capabilities for data and control. A later development was to grow hundreds of such nanowires on a substrate which functioned as an electrode. On top of this was placed a silicon electrode covered with a series of platinum ridges. Vibration of the top electrode caused the generation of direct current. A report by Wang was to appear in the August 8, 2007 issue of the journal "Nano letters," saying that such devices could power implantable biomedical devices. The device would be powered by flowing blood or a beating heart. It could function while immersed in body fluids, and would get its energy from ultrasonic vibrations. Wang expects that an array of the devices could produce 4 watts per cubic centimetre. Goals for further development are to increase the efficiency of the array of nanowires, and to increase the lifetime of the device, which as of April 2007 was only about one hour. By November 2010 Wang and his team were able to produce 3 volts of potential and as much as 300 nanoamperes of current, an output level 100 times greater than was possible a year earlier, from an array measuring about 2 cm by 1.5 cm.

The windbelt is a micropower technology invented by Shawn Frayne. It is essentially an aeolian harp except that it exploits the motion of the string produced by aeroelastic flutter to create a physical oscillation that can be converted to electricity. It avoids the losses inherent in rotating wind powered generators. Prototypes have produced 40 milliwatts in a 10 mile per hour wind. Magnets on the vibrating membrane generate currents in stationary coils.

Piezoelectric nanofibres in clothing could generate enough electricity from the wearer's body movements to power small electronic devices such as iPods or some of the electronic equipment used by soldiers on the battlefield, based on research by UC Berkeley Professor Liwei Lin and his team. One million such fibres could power an iPod, and would be altogether as large as a grain of sand. Researchers at Stanford University are developing "eTextiles," which are batteries made of fabric, which might serve to store power generated by such technology.

Microelectromechanical Systems

Microelectromechanical systems (MEMS) (also written as *micro-electro-mechanical*, *MicroElectroMechanical* or *microelectronic and microelectromechanical systems* and the related *micromechatronics*) is the technology of very small devices; it merges at the nano-scale into nanoelectromechanical systems (NEMS) and nanotechnology. MEMS are also referred to as micromachines (in Japan), or *micro systems technology – MST* (in Europe).

MEMS are separate and distinct from the hypothetical vision of molecular nanotechnology or molecular electronics. MEMS are made up of components between 1 to 100 micrometres in size (i.e. 0.001 to 0.1 mm), and MEMS devices generally range in size from 20 micrometres (20 millionths of a metre) to a millimetre (i.e. 0.02 to 1.0 mm). They usually consist of a central unit that processes data (the microprocessor) and several components that interact with the surroundings such as microsensors. At these size scales, the standard constructs of classical physics are not always useful. Because of the large surface area to volume ratio of MEMS, surface effects such as electrostatics and wetting dominate over volume effects such as inertia or thermal mass.

The potential of very small machines was appreciated before the technology existed that could make them—see, for example, Richard Feynman's famous 1959 lecture There's Plenty of Room at the Bottom. MEMS became practical once they could be fabricated using modified semiconductor device fabrication technologies, normally used to make electronics. These include molding and plating, wet etching (KOH, TMAH) and dry etching (RIE and DRIE), electro discharge machining (EDM), and other technologies capable of manufacturing small devices. An early example of a MEMS device is the resonistor – an electromechanical monolithic resonator.

Materials for MEMS Manufacturing

The fabrication of MEMS evolved from the process technology in semiconductor device fabrication, i.e. the basic techniques are deposition of material layers, patterning by photolithography and etching to produce the required shapes.

Silicon

Silicon is the material used to create most integrated circuits used in consumer electronics in the modern industry. The economies of scale, ready availability of cheap high-quality materials and ability to incorporate electronic functionality make silicon attractive for a wide variety of MEMS applications. Silicon also has significant advantages engendered through its material properties. In single crystal form, silicon is an almost perfect Hookean material, meaning that when it is flexed there is virtually no hysteresis and hence almost no energy dissipation. As well as making for highly repeatable motion, this also makes silicon very reliable as it suffers very little fatigue and can have service lifetimes in the range of billions to trillions of cycles without breaking.

Polymers

Even though the electronics industry provides an economy of scale for the silicon industry, crystalline silicon is still a complex and relatively expensive material to be produced. Polymers on the other hand can be produced in huge volumes, with a great variety of material characteristics. MEMS devices can be made from polymers by processes such as injection molding, embossing or stereolithography and are especially well suited to microfluidic applications such as disposable blood testing cartridges.

Metals

Metals can also be used to create MEMS elements. While metals do not have some of the advantages displayed by silicon in terms of mechanical properties, when used within their limitations, metals can exhibit very high degrees of reliability. Metals can be deposited by electroplating, evaporation, and sputtering processes. Commonly used metals include gold, nickel, aluminium, copper, chromium, titanium, tungsten, platinum, and silver.

Ceramics

The nitrides of silicon, aluminium and titanium as well as silicon carbide and other ceramics are increasingly applied in MEMS fabrication due to advantageous combinations of material properties. AlN crystallizes in the wurtzite structure and thus shows pyroelectric and piezoelectric properties enabling sensors, for instance, with sensitivity to normal and shear forces. TiN, on the other hand, exhibits a high electrical conductivity and large elastic modulus allowing to realise electrostatic MEMS actuation schemes with ultrathin membranes. Moreover, the high resistance of TiN against biocorrosion qualifies the material for applications in biogenic environments and in biosensors.

MEMS Basic Processes

Deposition Processes

One of the basic building blocks in MEMS processing is the ability to deposit thin films of material with a thickness anywhere between a few nanometres to about 100 micrometres. There are two types of deposition processes, as follows.

Physical Deposition

Physical vapor deposition ("PVD") consists of a process in which a material is removed from a target, and deposited on a surface. Techniques to do this include the process of sputtering, in which an ion beam liberates atoms from a target, allowing them to move through the intervening space and deposit on the desired substrate, and Evaporation (deposition), in which a material is evaporated from a target using either heat (thermal evaporation) or an electron beam (e-beam evaporation) in a vacuum system.

Chemical Deposition

Chemical deposition techniques include chemical vapor deposition ("CVD"), in which a stream of source gas reacts on the substrate to grow the material desired. This can be further divided into categories depending on the details of the technique, for example, LPCVD (Low Pressure chemical vapor deposition) and PECVD (Plasma Enhanced chemical vapor deposition).

Oxide films can also be grown by the technique of thermal oxidation, in which the (typically silicon) wafer is exposed to oxygen and/or steam, to grow a thin surface layer of silicon dioxide.

Patterning

Patterning in MEMS is the transfer of a pattern into a material.

Lithography

Lithography in MEMS context is typically the transfer of a pattern into a photosensitive material by selective exposure to a radiation source such as light. A photosensitive material is a material that experiences a change in its physical properties when exposed to a radiation source. If a photosensitive material is selectively exposed to radiation (e.g. by masking some of the radiation) the pattern of the radiation on the material is transferred to the material exposed, as the properties of the exposed and unexposed regions differs.

This exposed region can then be removed or treated providing a mask for the underlying substrate. Photolithography is typically used with metal or other thin film deposition, wet and dry etching.

Photolithography

Photolithography is the process of transferring geometric shapes on a mask to the surface of a silicon wafer. The steps involved in the photolithographic process are wafer cleaning; barrier layer formation; photoresist application; soft baking; mask alignment; exposure and development; and hard-baking.

Wafer Cleaning, Barrier Formation and Photoresist Application

In the first step, the wafers are chemically cleaned to remove particulate matter on the surface as well as any traces of organic, ionic, and metallic impurities. After cleaning, silicon dioxide, which serves as a barrier layer, is deposited on the surface of the wafer. After the formation of the SiO2 layer, photoresist is applied to the surface of the wafer. High-speed centrifugal whirling of silicon wafers is the standard method for applying photoresist coatings in IC manufacturing. This technique, known as "Spin Coating," produces a thin uniform layer of photoresist on the wafer surface. Another method for applying photoresist coatings onto MEMs wafers with high aspect ratio trenches uses an ultrasonic nozzle to spray micron-sized atomized droplets of photoresist directly onto the wafer surface.

Positive and Negative Photoresist: There are two types of photoresist: positive and negative. For positive resists, the resist is exposed with UV light wherever the underlying material is to be removed. In these resists, exposure to the UV light changes the chemical structure of the resist so that it becomes more soluble in the developer. The exposed resist is then washed away by the developer solution, leaving windows of the bare underlying material. In other words, "whatever shows, goes." The mask, therefore, contains an exact copy of the pattern which is to remain on the wafer.

Negative resists behave in just the opposite manner. Exposure to the UV light causes the negative resist to become polymerized, and more difficult to dissolve. Therefore, the negative resist remains on the surface wherever it is exposed, and the developer solution removes only the unexposed portions. Masks used for negative photoresists, therefore, contain the inverse (or photographic "negative") of the pattern to be transferred. The figure below shows the pattern differences generated from the use of positive and negative resist.

Electron Beam Lithography

Electron beam lithography (often abbreviated as e-beam lithography) is the practice of scanning a beam of electrons in a patterned fashion across a surface covered with a film (called the resist), ("exposing" the resist) and of selectively removing either exposed or non-exposed regions of the resist ("developing"). The purpose, as with photolithography, is to create very small structures in the resist that can subsequently be transferred to the substrate material, often by etching. It was developed for manufacturing integrated circuits, and is also used for creating nanotechnology architectures.

The primary advantage of electron beam lithography is that it is one of the ways to beat the diffraction limit of light and make features in the nanometre regime. This form of maskless lithography has found wide usage in photomask-making used in photolithography, low-volume production of semiconductor components, and research & development.

The key limitation of electron beam lithography is throughput, i.e., the very long time it takes to expose an entire silicon wafer or glass substrate. A long exposure time leaves the user vulnerable to beam drift or instability which may occur during the exposure. Also, the turn-around time for reworking or re-design is lengthened unnecessarily if the pattern is not being changed the second time.

Ion Beam Lithography

It is known that focused-ion-beam lithography has the capability of writing extremely fine lines (less than 50 nm line and space has been achieved) without proximity effect. However, because the writing field in ion-beam lithography is quite small, large area patterns must be created by stitching together the small fields.

Ion Track Technology

Ion track technology is a deep cutting tool with a resolution limit around 8 nm applicable to radiation resistant minerals, glasses and polymers. It is capable to generate holes in thin films without any development process. Structural depth can be defined either by ion range or by material thickness. Aspect ratios up to several 10^4 can be reached. The technique can shape and texture materials at a defined inclination angle. Random pattern, single-ion track structures and aimed pattern consisting of individual single tracks can be generated.

X-Ray Lithography

X-ray lithography, is a process used in electronic industry to selectively remove parts of a thin film. It uses X-rays to transfer a geometric pattern from a mask to a light-sensitive chemical photoresist, or simply "resist," on the substrate. A series of chemical treatments then engraves the produced pattern into the material underneath the photoresist.

Diamond Patterning

Etching Processes: There are two basic categories of etching processes: wet etching and dry etching. In the former, the material is dissolved when immersed in a chemical solution. In the latter, the material is sputtered or dissolved using reactive ions or a vapor phase etchant. for a somewhat dated overview of MEMS etching technologies.

Wet Etching

Wet chemical etching consists in selective removal of material by dipping a substrate into a solution that dissolves it. The chemical nature of this etching process provides a good selectivity, which means the etching rate of the target material is considerably higher than the mask material if selected carefully.

Isotropic Etching

Etching progresses at the same speed in all directions. Long and narrow holes in a mask will produce v-shaped grooves in the silicon. The surface of these grooves can be atomically smooth if the etch is carried out correctly, with dimensions and angles being extremely accurate.

Anisotropic Etching

Some single crystal materials, such as silicon, will have different etching rates depending on the crystallographic orientation of the substrate. This is known as anisotropic etching and one of the most common examples is the etching of silicon in KOH (potassium hydroxide), where Si <111> planes etch approximately 100 times slower than other planes (crystallographic orientations). Therefore, etching a rectangular hole in a (100)-Si wafer results in a pyramid shaped etch pit with 54.7° walls, instead of a hole with curved sidewalls as with isotropic etching.

HF Etching

Hydrofluoric acid is commonly used as an aqueous etchant for silicon dioxide (SiO_2, also known as BOX for SOI), usually in 49% concentrated form, 5:1, 10:1 or 20:1 BOE (buffered oxide etchant) or BHF (Buffered HF). They were first used in medieval times for glass etching. It was used in IC fabrication for patterning the gate oxide until the process step was replaced by RIE.

Hydrofluoric acid is considered one of the more dangerous acids in the cleanroom. It penetrates the skin upon contact and it diffuses straight to the bone. Therefore the damage is not felt until it is too late.

Electrochemical Etching

Electrochemical etching (ECE) for dopant-selective removal of silicon is a common method to automate and to selectively control etching. An active p-n diode junction is required, and either type of dopant can be the etch-resistant ("etch-stop") material. Boron is the most common etch-stop dopant. In combination with wet anisotropic etching as described above, ECE has been used successfully for controlling silicon diaphragm thickness in commercial piezoresistive silicon pressure sensors. Selectively doped regions can be created either by implantation, diffusion, or epitaxial deposition of silicon.

Dry Etching

Xenon Difluoride: Xenon difluoride (XeF_2) is a dry vapor phase isotropic etch for silicon originally applied for MEMS in 1995 at University of California, Los Angeles. Primarily used for releasing metal and dielectric structures by undercutting silicon, XeF_2 has the advantage of a stiction-free release unlike wet etchants. Its etch selectivity to silicon is very high, allowing it to work with photoresist, SiO_2, silicon nitride, and various metals for masking. Its reaction to silicon is "plasmaless", is purely chemical and spontaneous and is often operated in pulsed mode. Models of the etching action are available, and university laboratories and various commercial tools offer solutions using this approach.

Hydrogen Fluoride (HF)

Plasma Etching: Modern VLSI processes avoid wet etching, and use plasma etching instead. Plasma etchers can operate in several modes by adjusting the parameters of the plasma. Ordinary plasma etching operates between 0.1 and 5 Torr. (This unit of pressure, commonly used in vacuum engineering, equals approximately 133.3 pascals.) The plasma produces energetic free radicals, neutrally charged, that react at the surface of the wafer. Since neutral particles attack the wafer from all angles, this process is isotropic.

Plasma etching can be isotropic, i.e., exhibiting a lateral undercut rate on a patterned surface approximately the same as its downward etch rate, or can be anisotropic, i.e., exhibiting a smaller lateral undercut rate than its downward etch rate. Such anisotropy is maximized in deep reactive ion etching. The use of the term anisotropy for plasma etching should not be conflated with the use of the same term when referring to orientation-dependent etching.

The source gas for the plasma usually contains small molecules rich in chlorine or fluorine. For instance, carbon tetrachloride (CCl4) etches silicon and aluminium, and trifluoromethane etches silicon dioxide and silicon nitride. A plasma containing oxygen is used to oxidize ("ash") photoresist and facilitate its removal.

Ion milling, or sputter etching, uses lower pressures, often as low as 10-4 Torr (10 mPa). It bombards the wafer with energetic ions of noble gases, often Ar+, which knock atoms from the substrate by transferring momentum. Because the etching is performed by ions, which approach the wafer approximately from one direction, this process is highly anisotropic. On the other hand, it tends to display poor selectivity. Reactive-ion etching (RIE) operates under conditions intermediate between sputter and plasma etching (between 10–3 and 10–1 Torr). Deep reactive-ion etching (DRIE) modifies the RIE technique to produce deep, narrow features.

Sputtering

Reactive ion Etching (RIE): In reactive ion etching (RIE), the substrate is placed inside a reactor, and several gases are introduced. A plasma is struck in the gas mixture using an RF power source, which breaks the gas molecules into ions. The ions accelerate towards, and react with, the surface of the material being etched, forming another gaseous material. This is known as the chemical part of reactive ion etching. There is also a physical part, which is similar to the sputtering deposition process. If the ions have high enough energy, they can knock atoms out of the material to be etched without a chemical reaction. It is a very complex task to develop dry etch processes that balance chemical and physical etching, since there are many parameters to adjust. By changing the balance it is possible to influence the anisotropy of the etching, since the chemical part is isotropic and the physical part highly anisotropic the combination can form sidewalls that have shapes from rounded to vertical. RIE can be deep (Deep RIE or deep reactive ion etching (DRIE)).

Deep RIE (DRIE) is a special subclass of RIE that is growing in popularity. In this process, etch depths of hundreds of micrometres are achieved with almost vertical sidewalls. The primary technology is based on the so-called "Bosch process", named after the German company Robert Bosch, which filed the original patent, where two different gas compositions alternate in the reactor. Currently there are two variations of the DRIE. The first variation consists of three distinct steps (the Bosch Process as used in the Plasma-Therm tool) while the second variation only consists of two steps (ASE used in the STS tool).

In the 1st Variation, the etch cycle is as follows:

(i) SF6 isotropic etch;

(ii) C4F8 passivation;

(iii) SF6 anisoptropic etch for floor cleaning.

In the 2nd variation, steps (i) and (iii) are combined.

Both variations operate similarly. The C4F8 creates a polymer on the surface of the substrate, and the second gas composition (SF6 and O_2) etches the substrate. The polymer is immediately sputtered away by the physical part of the etching, but only on the horizontal surfaces and not the sidewalls. Since the polymer only dissolves very slowly in the chemical part of the etching, it builds up on the sidewalls and protects them from etching. As a result, etching aspect ratios of 50 to 1 can be achieved. The process can easily be used to etch completely through a silicon substrate, and etch rates are 3–6 times higher than wet etching.

Die Preparation

After preparing a large number of MEMS devices on a silicon wafer, individual dies have to be separated, which is called die preparation in semiconductor technology. For some applications, the separation is preceded by wafer backgrinding in order to reduce the wafer thickness. Wafer dicing may then be performed either by sawing using a cooling liquid or a dry laser process called stealth dicing.

MEMS Manufacturing Technologies

Bulk Micromachining

Bulk micromachining is the oldest paradigm of silicon based MEMS. The whole thickness of a silicon wafer is used for building the micro-mechanical structures. Silicon is machined using various etching processes. Anodic bonding of glass plates or additional silicon wafers is used for adding features in the third dimension and for hermetic encapsulation. Bulk micromachining has been essential in enabling high performance pressure sensors and accelerometres that changed the sensor industry in the 1980s and 90's.

Surface Micromachining

Surface micromachining uses layers deposited on the surface of a substrate as the structural materials, rather than using the substrate itself. Surface micromachining was created in the late 1980s to render micromachining of silicon more compatible with planar integrated circuit technology, with the goal of combining MEMS and integrated circuits on the same silicon wafer. The original surface micromachining concept was based on thin polycrystalline silicon layers patterned as movable mechanical structures and released by sacrificial etching of the underlying oxide layer. Interdigital comb electrodes were used to produce in-plane forces and to detect in-plane movement capacitively. This MEMS paradigm has enabled the manufacturing of low cost accelerometres for e.g. automotive air-bag systems and other applications where low performance and/or high g-ranges are sufficient. Analog Devices have pioneered the industrialization of surface micromachining and have realised the co-integration of MEMS and integrated circuits.

High Aspect Ratio (HAR) Silicon Micromachining

Both bulk and surface silicon micromachining are used in the industrial production of sensors, ink-jet nozzles, and other devices. But in many cases the distinction between these two has diminished. A new etching technology, deep reactive-ion etching, has made it possible to combine good performance typical of bulk micromachining with comb structures and in-plane operation typical of surface micromachining. While it is common in surface micromachining to have structural layer thickness in the range of 2 μm, in HAR silicon micromachining the thickness can be from 10 to 100 μm. The materials commonly used in HAR silicon micromachining are thick polycrystalline silicon, known as epi-poly, and bonded silicon-on-insulator (SOI) wafers although processes for bulk silicon wafer also have been created (SCREAM). Bonding a second wafer by glass frit bonding, anodic bonding or alloy bonding is used to protect the MEMS structures. Integrated circuits are typically not combined with HAR silicon micromachining.

Applications

Figure: *microelectromechanical systems chip, sometimes called "lab on a chip"*

Figure: *MEMS compared to a 0.02 € coin*

In one viewpoint MEMS application is categorized by type of use.

- Sensor
- Actuator
- Structure

In another view point MEMS applications are categorized by the field of application (commercial applications include):

- Inkjet printers, which use piezoelectrics or thermal bubble ejection to deposit ink on paper.
- Accelerometres in modern cars for a large number of purposes including airbag deployment in collisions.
- Accelerometres in consumer electronics devices such as game controllers (Nintendo Wii), personal media players / cell phones (Apple iPhone, various Nokia mobile phone models, various HTC PDA models) and a number of Digital Cameras (various Canon Digital IXUS models). Also used in PCs to park the hard disk head when free-fall is detected, to prevent damage and data loss.
- MEMS gyroscopes used in modern cars and other applications to detect yaw; e.g., to deploy a roll over bar or trigger dynamic stability control
- MEMS microphones in portable devices, e.g., mobile phones, head sets and laptops.
- Silicon pressure sensors e.g., car tire pressure sensors, and disposable blood pressure sensors
- Displays e.g., the DMD chip in a projector based on DLP technology, which has a surface with several hundred thousand micromirrors or single micro-scanning-mirrors also called microscanners
- Optical switching technology, which is used for switching technology and alignment for data communications
- Bio-MEMS applications in medical and health related technologies from Lab-On-Chip to MicroTotalAnalysis (biosensor, chemosensor), or embedded in medical devices e.g. stents.
- Interferometric modulator display (IMOD) applications in consumer electronics (primarily displays for mobile devices), used to create interferometric modulation " reflective display technology as found in mirasol displays
- Fluid acceleration such as for micro-cooling
- Micro-scale Energy harvesting including piezoelectric, electrostatic and electromagnetic micro harvesters.

- Micromachined Ultrasound Transducer including Piezoelectric Micromachined Ultrasonic Transducers and Capacitive Micromachined Ultrasonic Transducers.

Companies with strong MEMS programs come in many sizes. The larger firms specialise in manufacturing high volume inexpensive components or packaged solutions for end markets such as automobiles, biomedical, and electronics. The successful small firms provide value in innovative solutions and absorb the expense of custom fabrication with high sales margins. In addition, both large and small companies work in R&D to explore MEMS technology.

Industry Structure

The global market for micro-electromechanical systems, which includes products such as automobile airbag systems, display systems and inkjet cartridges totaled $40 billion in 2006 according to Global MEMS/Microsystems Markets and Opportunities, a research report from SEMI and Yole Development and is forecasted to reach $72 billion by 2011.

MEMS devices are defined as die-level components of first-level packaging, and include pressure sensors, accelerometres, gyroscopes, microphones, digital mirror displays, microfluidic devices, etc. The materials and equipment used to manufacture MEMS devices topped $1 billion worldwide in 2006. Materials demand is driven by substrates, making up over 70 percent of the market, packaging coatings and increasing use of chemical mechanical planarization (CMP). While MEMS manufacturing continues to be dominated by used semiconductor equipment, there is a migration to 200 mm lines and select new tools, including etch and bonding for certain MEMS applications.

4

Nanoelectromechanical Systems

Nanoelectromechanical systems (NEMS) are a class of devices integrating electrical and mechanical functionality on the nanoscale. NEMS form the logical next miniaturization step from so-called microelectromechanical systems, or MEMS devices. NEMS typically integrate transistor-like nanoelectronics with mechanical actuators, pumps, or motors, and may thereby form physical, biological, and chemical sensors. The name derives from typical device dimensions in the nanometre range, leading to low mass, high mechanical resonance frequencies, potentially large quantum mechanical effects such as zero point motion, and a high surface-to-volume ratio useful for surface-based sensing mechanisms. Uses include accelerometres, or detectors of chemical substances in the air.

Because of the scale on which they can function, NEMS are expected to significantly impact many areas of technology and science and eventually replace MEMS. As noted by Richard Feynman in his famous talk in 1959, "There's Plenty of Room at the Bottom," there are a lot of potential applications of machines at smaller and smaller sizes; by building and controlling devices at smaller scales, all technology benefits. Among the expected benefits include greater efficiencies and reduced size, decreased power consumption and lower costs of production in electromechanical systems.

In 2000, the first very-large-scale integration (VLSI) NEMS device was demonstrated by researchers from IBM. Its premise was an array of AFM tips which can heat/sense a deformable substrate in order to function as a memory device. In 2007, the International Technical Roadmap for Semiconductors (ITRS) contains NEMS Memory as a new entry for the Emerging Research Devices section.

Importance for AFM

A key application of NEMS is atomic force microscope tips. The increased sensitivity achieved by NEMS leads to smaller and more efficient sensors to detect stresses, vibrations, forces at the atomic level, and chemical signals. AFM tips and other detection at the nanoscale rely heavily on NEMS. If implementation of better scanning devices becomes available, all of nanoscience could benefit from AFM tips.

Approaches to Miniaturization

Two complementary approaches to fabrication of NEMS can be found. The top-down approach uses the traditional microfabrication methods, i.e. optical and electron beam lithography, to manufacture devices. While being limited by the resolution of these methods, it allows a large degree of control over the resulting structures. Typically, devices are fabricated from metallic thin films or etched semiconductor layers.

Bottom-up approaches, in contrast, use the chemical properties of single molecules to cause single-molecule components to (a) self-organise or self-assemble into some useful conformation, or (b) rely on positional assembly. These approaches utilise the concepts of molecular self-assembly and/or molecular recognition. This allows fabrication of much smaller structures, albeit often at the cost of limited control of the fabrication process.

A combination of these approaches may also be used, in which nanoscale molecules are integrated into a top-down framework. One such example is the carbon Nanotube nanomotor.

Materials

Carbon Allotropes

Many of the commonly used materials for NEMS technology have been carbon based, specifically diamond, carbon nanotubes and graphene. This is mainly because of the useful properties of carbon based materials which directly meet the needs of NEMS. The mechanical properties of carbon (such as large Young's modulus) are fundamental to the stability of NEMS while the metallic and semiconductor conductivities of carbon based materials allow them to function as transistors.

Both graphene and diamond exhibit high Young's modulus, low density, low friction, excessively low mechanical dissipation, and large surface area. The low friction of CNTs, allow practically frictionless bearings and has thus been a huge motivation towards practical applications of CNTs as constitutive elements in NEMS, such as nanomotors, switches, and high-frequency oscillators. Carbon nanotubes and graphene's physical strength allows carbon based materials to meet higher stress demands, when common materials would normally fail and thus further support their use as a major materials in NEMS technological development.

Along with the mechanical benefits of carbon based materials, the electrical properties of carbon nanotubes and graphene allow it to be used in many electrical components of NEMS. Nanotransistors have been developed for both carbon nanotubes as well as graphene. Transistors are one of the basic building blocks for all electronic devices, so by effectively developing usable transistors, carbon nanotubes and graphene are both very crucial to NEMS.

Metallic Carbon Nanotubes

Metallic carbon nanotubes have also been proposed for nanoelectronic interconnects since they can carry high current densities. This is a very useful property as wires to transfer current are another basic building block of any electrical system. Carbon nanotubes have specifically found so much use in NEMS that methods have already been discovered to connect suspended carbon nanotubes to other nanostructures. This allows carbon nanotubes to be structurally set up to make complicated nanoelectric systems. Because carbon based products can be properly controlled and act as interconnects as well as transistors, they serve as a fundamental material in the electrical components of NEMS.

Difficulties

Despite all of the useful properties of carbon nanotubes and graphene for NEMS technology, both of these products face several hindrances to their implementation. One of the main problems is carbon's response to real life environments. Carbon nanotubes exhibit a large change in electronic properties when exposed to oxygen. Similarly, other changes to the electronic and mechanical attributes of carbon based materials must fully be explored before their implementation, especially because of their high surface area which can easily react with surrounding environments. Carbon Nanotubes were also found to have varying conductivities, being either metallic or semiconducting depending on their helicity when processed.

Because of this, very special treatment must be given to the nanotubes during processing, in order to assure that all of the nanotubes have appropriate conductivities. Graphene also has very complicated electric conductivity properties compared to traditional semiconductors as it lacks an energy band gap and essentially changes all the rules for how electrons move through a graphene based device. This means that traditional constructions of electronic devices will likely not work and completely new architectures must be designed for these new electronic devices.

Simulations

Computer simulations have long been important counterparts to experimental studies of NEMS devices. Through continuum mechanics and molecular dynamics (MD), important behaviours of NEMS devices can be predicted via computational modelling before engaging in experiments. Additionally, combining continuum and MD techniques enables engineers to efficiently analyze the stability of NEMS devices without resorting to ultra-fine meshes and time-intensive simulations. Simulations have other advantages as well: they do not require the time and expertise associated with fabricating NEMS devices; they can effectively predict the interrelated roles of various electromechanical effects; and parametric studies can be conducted fairly readily as compared with experimental approaches. For example, computational studies have predicted the charge distributions and "pull-in" electromechanical responses of NEMS devices. Using simulations to predict mechanical and electrical behaviour of these devices can help optimize NEMS device design parameters.

Future of NEMS

Key hurdles currently preventing the commercial application of many NEMS devices include low-yields and high device quality variability. Before NEMS devices can actually be implemented, reasonable integrations of carbon based products must be created. A recent step in that direction has been demonstrated for diamond, achieving a processing level comparable to that of silicon. The focus is currently shifting from experimental work towards practical applications and device structures that will implement and profit from of such novel devices. The next challenge to overcome involves understanding all of the properties of these carbon-based tools, and using the properties to make efficient and durable NEMS with low failure rates.

NEMS devices, if implemented into everyday technologies, could further reduce the size of modern devices and allow for better performing sensors. Carbon-based materials have served as prime materials for NEMS use, because of their exceptional mechanical and electrical properties. Once NEMS interactions with outside environments are integrated with effective designs, they will likely become useful products to everyday technologies.

Global Market of NEMS projected to reach $108.88 million by 2022.

Battery (Electricity)

An electric battery is a device consisting of one or more electrochemical cells that convert stored chemical energy into electrical energy. Each cell contains a positive terminal, or cathode, and a negative terminal, or anode. Electrolytes allow ions to move between the electrodes and terminals, which allows current to flow out of the battery to perform work.

Primary (single-use or "disposable") batteries are used once and discarded; the electrode materials are irreversibly changed during discharge. Common examples are the alkaline battery used for flashlights and a multitude of portable devices. Secondary (rechargeable batteries) can be discharged and recharged multiple times; the original composition of the electrodes can be restored by reverse current. Examples include the lead-acid batteries used in vehicles and lithium ion batteries used for portable electronics. Batteries come in many shapes and sizes, from miniature cells used to power hearing aids and wristwatches to battery banks the size of rooms that provide standby power for telephone exchanges and computer data centres.

According to a 2005 estimate, the worldwide battery industry generates US$48 billion in sales each year, with 6% annual growth.

Batteries have much lower specific energy (energy per unit mass) than common fuels such as gasoline. This is somewhat mitigated by the fact that batteries deliver their energy as electricity (which can be converted efficiently to mechanical work), whereas using fuels in engines entails a low efficiency of conversion to work.

The usage of "battery" to describe a group electrical devices dates to Benjamin Franklin, who in 1748 described multiple Leyden jars by analogy to a battery of cannon (Benjamin Franklin borrowed the term "battery" from the military, which refers to weapons functioning together). Alessandro Volta described the first electrochemical battery, the voltaic pile in 1800. This was a stack of copper and zinc plates, separated by brine soaked paper disks, that could produce a steady current for a considerable length of time. Volta did not appreciate that the voltage was due to chemical reactions. He thought that his cells were an inexhaustible source of energy, and that the associated corrosion effects at the electrodes were a mere nuisance, rather than an unavoidable consequence of their operation, as Michael Faraday showed in 1834.

Although early batteries were of great value for experimental purposes, in practice their voltages fluctuated and they could not provide a large current for a sustained period. The Daniell cell, invented in 1836 by British chemist John Frederic Daniell, was the first practical source of electricity, becoming an industry standard and seeing widespread adoption as a power source for electrical telegraph networks. It consisted of a copper pot filled with a copper sulfate solution, in which was immersed an unglazed earthenware container filled with sulfuric acid and a zinc electrode.

These wet cells used liquid electrolytes, which were prone to leakage and spillage if not handled correctly. Many used glass jars to hold their components, which made them fragile. These characteristics made wet cells unsuitable for portable appliances. Near the end of the nineteenth century, the invention of dry cell batteries, which replaced the liquid electrolyte with a paste, made portable electrical devices practical.

Principle of Operation

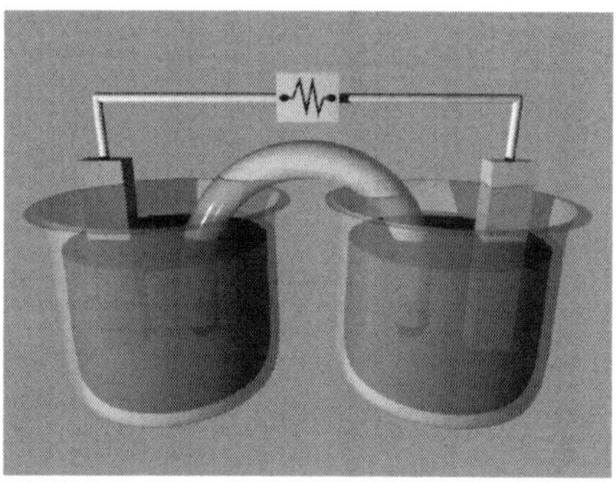

Figure: *A voltaic cell for demonstration purposes. In this example the two half-cells are linked by a salt bridge separator that permits the transfer of ions.*

Batteries convert chemical energy directly to electrical energy. A battery consists of some number of voltaic cells. Each cell consists of two half-cells connected in series by a conductive electrolyte containing anions and cations. One half-cell includes electrolyte and the negative electrode, the electrode to which anions (negatively charged ions) migrate; the other half-cell includes electrolyte and the positive electrode to which cations (positively charged ions) migrate. Redox reactions power the battery. Cations are reduced (electrons are added) at the cathode during charging, while anions are oxidized (electrons are removed) at the anode during discharge. The electrodes do not touch each other, but are electrically connected by the electrolyte. Some cells use different electrolytes for each half-cell. A separator allows ions to flow between half-cells, but prevents mixing of the electrolytes.

Each half-cell has an electromotive force (or emf), determined by its ability to drive electric current from the interior to the exterior of the cell. The net emf of the cell is the difference between the emfs of its half-cells. Thus, if the electrodes have emfs $\mathcal{E}_1$ and $\mathcal{E}_2$, then the net emf is $\mathcal{E}_2 - \mathcal{E}_1$; in other words, the net emf is the difference between the reduction potentials of the half-reactions.

The electrical driving force or ΔV_{bat} across the terminals of a cell is known as the *terminal voltage (difference)* and is measured in volts. The terminal voltage of a cell that is neither charging nor discharging is called the open-circuit voltage and equals the emf of the cell. Because of internal resistance, the terminal voltage of a cell that is discharging is smaller in magnitude than the open-circuit voltage and the terminal voltage of a cell that is charging exceeds the open-circuit voltage. An ideal cell has negligible internal resistance, so it would maintain a constant terminal voltage of ε until exhausted, then dropping to zero. If such a cell maintained 1.5 volts and stored a charge of one coulomb then on complete discharge it would perform 1.5 joules of work. In actual cells, the internal resistance increases under discharge and the open circuit voltage also decreases under discharge. If the voltage and resistance are plotted against time, the resulting graphs typically are a curve; the shape of the curve varies according to the chemistry and internal arrangement employed.

The voltage developed across a cell's terminals depends on the energy release of the chemical reactions of its electrodes and electrolyte. Alkaline and zinc–carbon cells have different chemistries, but approximately the same emf of 1.5 volts; likewise NiCd and NiMH cells have different chemistries, but approximately the same emf of 1.2 volts. The high electrochemical potential changes in the reactions of lithium compounds give lithium cells emfs of 3 volts or more.

Categories and Types of Batteries

Batteries are classified into primary and secondary forms.

- *Primary* batteries irreversibly transform chemical energy to electrical energy. When the supply of reactants is exhausted, energy cannot be readily restored to the battery.

- *Secondary* batteries can be recharged; that is, they can have their chemical reactions reversed by supplying electrical energy to the cell, approximately restoring their original composition.

Some types of primary batteries used, for example, for telegraph circuits, were restored to operation by replacing the electrodes. Secondary batteries are not indefinitely rechargeable due to dissipation of the active materials, loss of electrolyte and internal corrosion.

Primary Batteries

Primary batteries, or primary cells, can produce current immediately on assembly. These are most commonly used in portable devices that have low current drain, are used only intermittently, or are used well away from an alternative power source, such as in alarm and communication circuits where other electric power is only intermittently available. Disposable primary cells cannot be reliably recharged, since the chemical reactions are not easily reversible and active materials may not return to their original forms. Battery manufacturers recommend against attempting to recharge primary cells.

In general, these have higher energy densities than rechargeable batteries, but disposable batteries do not fare well under high-drain applications with loads under 75 ohms (75 Ω).

Common types of disposable batteries include zinc–carbon batteries and alkaline batteries.

Secondary Batteries

Secondary batteries, also known as *secondary cells*, or *rechargeable batteries*, must be charged before first use; they are usually assembled with active materials in the discharged state. Rechargeable batteries are (re)charged by applying electric current, which reverses the chemical reactions that occur during discharge/use. Devices to supply the appropriate current are called chargers.

The oldest form of rechargeable battery is the lead–acid battery. This technology contains liquid electrolyte in an unsealed container, requiring that the battery be kept upright and the area be well ventilated to ensure safe dispersal of the hydrogen gas it produces during overcharging. The lead–acid battery is relatively heavy for the amount of electrical energy it can supply. Its low manufacturing cost and its high surge current levels make it common where its capacity (over approximately 10 Ah) is more important than weight and handling issues. A common application is the modern car battery, which can, in general, deliver a peak current of 450 amperes.

The sealed valve regulated lead–acid battery (VRLA battery) is popular in the automotive industry as a replacement for the lead–acid wet cell. The VRLA battery uses an immobilized sulfuric acid electrolyte, reducing the chance of leakage and extending shelf life. VRLA batteries immobilize the electrolyte. The two types are:

- *Gel batteries* (or "gel cell") use a semi-solid electrolyte.
- *Absorbed Glass Mat* (AGM) batteries absorb the electrolyte in a special fibreglass matting.

Other portable rechargeable batteries include several sealed "dry cell" types, that are useful in applications such as mobile phones and laptop computers. Cells of this type (in order of increasing power density and cost) include nickel–cadmium (NiCd), nickel–zinc (NiZn), nickel metal hydride (NiMH), and lithium-ion (Li-ion) cells. Li-ion has by far the highest share of the dry cell rechargeable market. NiMH has replaced NiCd in most applications due to its higher capacity, but NiCd remains in use in power tools, two-way radios, and medical equipment.

Recent developments include batteries with embedded electronics such as USBCELL, which allows charging an AA battery through a USB connector, and smart battery packs with state-of-charge monitors and battery protection circuits that prevent damage on over-discharge. Low self-discharge (LSD) allows secondary cells to be charged prior to shipping.

Battery Cell Types

Many types of electrochemical cells have been produced, with varying chemical processes and designs, including galvanic cells, electrolytic cells, fuel cells, flow cells and voltaic piles.

Wet Cell

A *wet cell* battery has a liquid electrolyte. Other names are *flooded cell*, since the liquid covers all internal parts, or *vented cell*, since gases produced during operation can escape to the air. Wet cells were a precursor to dry cells and are commonly used as a learning tool for electrochemistry. They can be built with common laboratory supplies, such as beakers, for demonstrations of how electrochemical cells work. A particular type of wet cell known as a concentration cell is important in understanding corrosion. Wet cells may be primary cells (non-rechargeable) or secondary cells (rechargeable). Originally, all practical primary batteries such as the Daniell cell were built as open-top glass jar wet cells. Other primary wet cells are the Leclanche cell, Grove cell,

Bunsen cell, Chromic acid cell, Clark cell, and Weston cell. The Leclanche cell chemistry was adapted to the first dry cells. Wet cells are still used in automobile batteries and in industry for standby power for switchgear, telecommunication or large uninterruptible power supplies, but in many places batteries with gel cells have been used instead. These applications commonly use lead–acid or nickel–cadmium cells.

Dry Cell

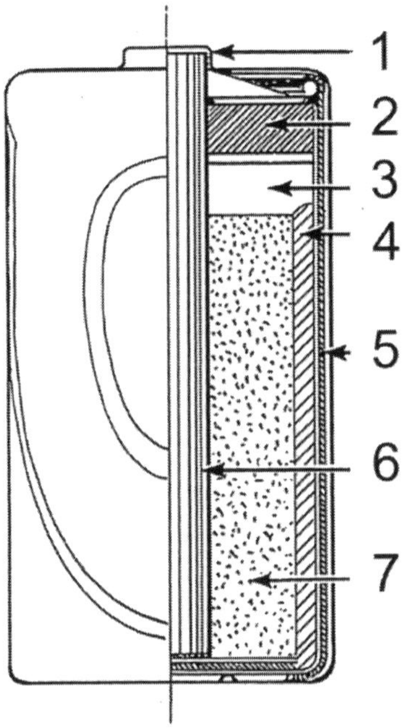

Figure: *Line art drawing of a dry cell: 1. brass cap, 2. plastic seal, 3. expansion space, 4. porous cardboard, 5. zinc can, 6. carbon rod, 7. chemical mixture.*

A *dry cell* uses a paste electrolyte, with only enough moisture to allow current to flow. Unlike a wet cell, a dry cell can operate in any orientation without spilling, as it contains no free liquid, making it suitable for portable equipment. By comparison, the first wet cells were typically fragile glass containers with lead rods hanging from the open top and needed careful handling to avoid spillage. Lead–acid batteries did not achieve the safety and portability of the dry cell until the development of the gel battery.

A common dry cell is the zinc–carbon battery, sometimes called the dry Leclanché cell, with a nominal voltage of 1.5 volts, the same as the alkaline battery (since both use the same zinc–manganese dioxide combination).

A standard dry cell comprises a zinc anode, usually in the form of a cylindrical pot, with a carbon cathode in the form of a central rod. The electrolyte is ammonium chloride in the form of a paste next to the zinc anode. The remaining space between the electrolyte and carbon cathode is taken up by a second paste consisting of ammonium chloride and manganese dioxide, the latter acting as a depolariser. In some designs, the ammonium chloride is replaced by zinc chloride.

Molten Salt

Molten salt batteries are primary or secondary batteries that use a molten salt as electrolyte. They operate at high temperatures and must be well insulated to retain heat.

Reserve

A reserve battery can be stored unassembled (unactivated and supplying no power) for a long period (perhaps years). When the battery is needed, then it is assembled (e.g., by adding electrolyte); once assembled, the battery is charged and ready to work. For example, a battery for an electronic artillery fuze might be activated by the impact of firing a gun: The acceleration breaks a capsule of electrolyte that activates the battery and powers the fuze's circuits. Reserve batteries are usually designed for a short service life (seconds or minutes) after long storage (years). A water-activated battery for oceanographic instruments or military applications becomes activated on immersion in water.

Battery Cell Performance

A battery's characteristics may vary over load cycle, over charge cycle, and over lifetime due to many factors including internal chemistry, current drain, and temperature.

Capacity and Discharge

A battery's *capacity* is the amount of electric charge it can deliver at the rated voltage. The more electrode material contained in the cell the greater its capacity. A small cell has less capacity than a larger cell with the same chemistry, although they develop the same open-circuit voltage. Capacity is measured in units such as amp-hour (A•h).

The rated capacity of a battery is usually expressed as the product of 20 hours multiplied by the current that a new battery can consistently supply for 20 hours at 68 °F (20 °C), while remaining above a specified terminal voltage per cell. For example, a battery rated at 100 A•h can deliver 5 A over a 20-hour period at room temperature.

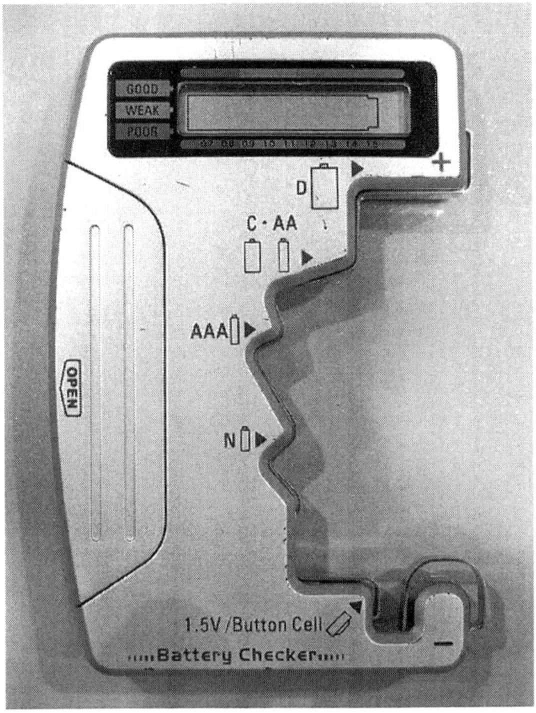

Figure: *A device to check battery voltage*

The fraction of the stored charge that a battery can deliver depends on multiple factors, including battery chemistry, the rate at which the charge is delivered (current), the required terminal voltage, the storage period, ambient temperature and other factors.

The higher the discharge rate, the lower the capacity. The relationship between current, discharge time and capacity for a lead acid battery is approximated (over a typical range of current values) by Peukert's law:

$$t = \frac{Q_P}{I^k}$$

where

Q_P is the capacity when discharged at a rate of 1 amp.

I is the current drawn from battery (A).

t is the amount of time (in hours) that a battery can sustain.

k is a constant around 1.3.

Batteries that are stored for a long period or that are discharged at a small fraction of the capacity lose capacity due to the presence of generally irreversible *side reactions* that consume charge carriers without producing current. This phenomenon is known as internal self-discharge. Further, when batteries are recharged, additional side reactions can occur, reducing capacity for subsequent discharges. After enough recharges, in essence all capacity is lost and the battery stops producing power.

Internal energy losses and limitations on the rate that ions pass through the electrolyte cause battery efficiency to vary. Above a minimum threshold, discharging at a low rate delivers more of the battery's capacity than at a higher rate. Installing batteries with varying A·h ratings does not affect device operation (although it may affect the operation interval) rated for a specific voltage unless load limits are exceeded. High-drain loads such as digital cameras can reduce total capacity, as happens with alkaline batteries. For example, a battery rated at 2000 mAh for a 10- or 20-hour discharge would not sustain a current of 1 A for a full two hours as its stated capacity implies.

C Rate

The C-rate is the multiple of the current over the current that a battery can sustain for one hour. A rate of 1 C means that an entire 1.6Ah battery would be discharged in one hour at a discharge current of 1.6 A. A 2C rate would mean a discharge current of 3.2 A, over one half-hour.

Fast-Charging, Large and Light Batteries

As of 2012 Lithium iron phosphate ($LiFePO_4$) battery technology was the fastest-charging/discharging, fully discharging in 10–20 seconds.

As of 2013, the world's largest battery was in Hebei Province, China. It stored 36 megawatt-hours of electricity at a cost of $500 million. Another large battery, composed of Ni–Cd cells, was in Fairbanks, Alaska. It covers 2,000 square metres (22,000 sq ft)—bigger than a football pitch—and weighs 1,300 tonnes, It was manufactured by ABB to provide backup power in the event of a blackout. The battery can provide 40 megawatts of power for up to seven minutes. Sodium–sulfur batteries have been used to store wind power. A 4.4 megawatt-hour battery system that can deliver 11 megawatts for 25 minutes stabilizes the output of the Auwahi wind farm in Hawaii. Lithium–sulfur batteries were used on the longest and highest solar-powered flight. The recharging speed of lithium-ion batteries can be increased by manufacturing changes.

Battery Lifetime

Available capacity of all batteries drops with decreasing temperature. In contrast to most of today's batteries, the Zamboni pile, invented in

1812, offers a very long service life without refurbishment or recharge, although it supplies current only in the nanoamp range. The Oxford Electric Bell has been ringing almost continuously since 1840 on its original pair of batteries, thought to be Zamboni piles.

Self-Discharge

Disposable batteries typically lose 8 to 20 percent of their original charge per year when stored at room temperature (20°–30 °C). This is known as the "self-discharge" rate, and is due to non-current-producing "side" chemical reactions that occur within the cell even when no load is applied. The rate of side reactions is reduced for batteries are stored at lower temperatures, although some can be damaged by freezing.

Old rechargeable batteries self-discharge more rapidly than disposable alkaline batteries, especially nickel-based batteries; a freshly charged nickel cadmium (NiCd) battery loses 10% of its charge in the first 24 hours, and thereafter discharges at a rate of about 10% a month. However, newer low self-discharge nickel metal hydride (NiMH) batteries and modern lithium designs display a lower self-discharge rate (but still higher than for primary batteries).

Corrosion

Internal parts may corrode and fail, or the active materials may be slowly converted to inactive forms.

Physical Component Changes

The active material on the battery plates changes chemical composition on each charge and discharge cycle, active material may be lost due to physical changes of volume; further limiting the number of times the battery can be recharged.

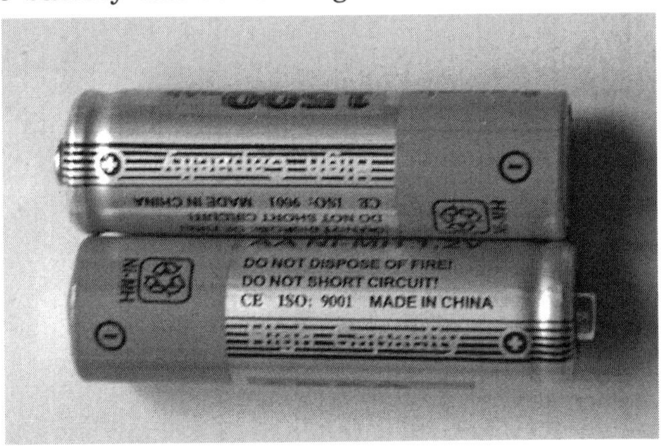

Figure: *Rechargeable batteries.*

Most nickel-based batteries are partially discharged when purchased, and must be charged before first use. Newer NiMH batteries are ready to be used when purchased, and have only 15% discharge in a year. Some deterioration occurs on each charge–discharge cycle. Degradation usually occurs because electrolyte migrates away from the electrodes or because active material detaches from the electrodes. Low-capacity NiMH batteries (1700–2000 mA·h) can be charged some 1,000 times, whereas high-capacity NiMH batteries (above 2500 mA·h) last about 500 cycles. NiCd batteries tend to be rated for 1,000 cycles before their internal resistance permanently increases beyond usable values.

Charge/Discharge Speed

Fast charging increases component changes, shortening battery lifespan.

Overcharging

If a charger cannot detect when the battery is fully charged then overcharging is likely, damaging it.

Memory Effect

NiCd cells, if used in a particular repetitive manner, may show a decrease in capacity called "memory effect". The effect can be avoided with simple practices. NiMH cells, although similar in chemistry, suffer less from memory effect.

Figure: *An analog camcorder battery [lithium ion].*

Environmental Conditions

Automotive lead–acid rechargeable batteries must endure stress due to vibration, shock, and temperature range. Because of these stresses and sulfation of their lead plates, few automotive batteries last beyond six years of regular use. Automotive starting (SLI: *Starting, Lighting, Ignition*) batteries have many thin plates to maximize current. In general, the thicker the plates the longer the life. They are typically discharged only slightly before recharge.

"Deep-cycle" lead–acid batteries such as those used in electric golf carts have much thicker plates to extend longevity. The main benefit of the lead–acid battery is its low cost; its main drawbacks are large size and weight for a given capacity and voltage. Lead–acid batteries should never be discharged to below 20% of their capacity, because internal resistance will cause heat and damage when they are recharged. Deep-cycle lead–acid systems often use a low-charge warning light or a low-charge power cut-off switch to prevent the type of damage that will shorten the battery's life.

Storage

Battery life can be extended by storing the batteries at a low temperature, as in a refrigerator or freezer, which slows the side reactions. Such storage can extend the life of alkaline batteries by about 5%; rechargeable batteries can hold their charge much longer, depending upon type. To reach their maximum voltage, batteries must be returned to room temperature; discharging an alkaline battery at 250 mA at 0 °C is only half as efficient as at 20 °C. Alkaline battery manufacturers such as Duracell do not recommend refrigerating batteries.

Battery Sizes

Primary batteries readily available to consumers range from tiny button cells used for electric watches, to the No. 6 cell used for signal circuits or other long duration applications. Secondary cells are made in very large sizes; very large batteries can power a submarine or stabilize an electrical grid and help level out peak loads.

Hazards

Explosion: A battery explosion is caused by misuse or malfunction, such as attempting to recharge a primary (non-rechargeable) battery, or a short circuit. Car batteries are most likely to explode when a short-circuit generates very large currents. Car batteries produce hydrogen, which is very explosive, when they are overcharged (because of electrolysis of the water in the electrolyte). The amount of overcharging is usually very small and generates little hydrogen, which dissipates quickly. However, when "jumping" a car battery, the high current can cause the rapid release of large volumes of hydrogen, which can be ignited explosively by a nearby spark, for example, when disconnecting a jumper cable.

When a battery is recharged at an excessive rate, an explosive gas mixture of hydrogen and oxygen may be produced faster than it can escape from within the battery, leading to pressure build-up and eventual bursting of the battery case. In extreme cases, battery acid may spray violently from the casing and cause injury. Overcharging—that is, attempting to charge a battery beyond its electrical capacity—can also lead to a battery explosion, in addition to leakage or irreversible damage. It may also cause damage to the charger or device in which the overcharged battery is later used. In addition, disposing of a battery via incineration may cause an explosion as steam builds up within the sealed case.

Leakage

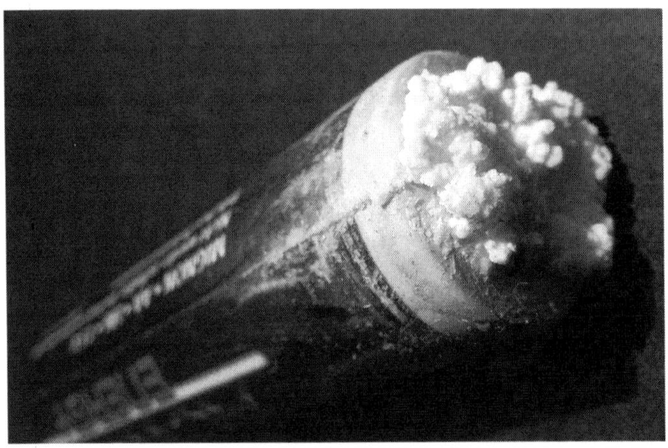

***Figure:** Leaked alkaline battery.*

Many battery chemicals are corrosive, poisonous or both. If leakage occurs, either spontaneously or through accident, the chemicals released may be dangerous.

For example, disposable batteries often use a zinc "can" both as a reactant and as the container to hold the other reagents. If this kind of battery is over-discharged, the reagents can emerge through the cardboard and plastic that form the remainder of the container. The active chemical leakage can then damage the equipment that the batteries power. For this reason, many electronic device manufacturers recommend removing the batteries from devices that will not be used for extended periods of time.

Toxic Materials

Many types of batteries employ toxic materials such as lead, mercury, and cadmium as an electrode or electrolyte. When each battery reaches end of life it must be disposed of to prevent environmental damage. Battery are one form of electronic waste (e-waste).

E-waste recycling services recover toxic substances, which can then be used for new batteries.

Of the nearly three billion batteries purchased annually in the United States, about 179,000 tons end up in landfills across the country.

In the United States, the Mercury-Containing and Rechargeable Battery Management Act of 1996 banned the sale of mercury-containing batteries, enacted uniform labelling requirements for rechargeable batteries and required that rechargeable batteries be easily removable. California and New York City prohibit the disposal of rechargeable batteries in solid waste, and along with Maine require recycling of cell phones. The rechargeable battery industry operates nationwide recycling programs in the United States and Canada, with dropoff points at local retailers.

The Battery Directive of the European Union has similar requirements, in addition to requiring increased recycling of batteries and promoting research on improved battery recycling methods.

In accordance with this directive all batteries to be sold within the EU must be marked with the "collection symbol" (A crossed-out wheeled bin). This must cover at least 3% of the surface of prismatic batteries and 1.5% of the surface of cylindrical batteries. All packaging must be marked likewise.

Ingestion

Batteries may be harmful or fatal if swallowed.

Small button cells can be swallowed, in particular by young children. While in the digestive tract, the battery's electrical discharge may lead to tissue damage; such damage is occasionally serious and can lead to death. Ingested disk batteries do not usually cause problems unless they become lodged in the gastrointestinal tract. The most common place for disk batteries to become lodged is the esophagus, resulting in clinical sequelae. Batteries that successfully traverse the esophagus are unlikely to lodge elsewhere. The likelihood that a disk battery will lodge in the esophagus is a function of the patient's age and battery size. Disk batteries of 16 mm have become lodged in the esophagi of 2 children younger than 1 year. Older children do not have problems with batteries smaller than 21–23 mm. Liquefaction necrosis may occur because sodium hydroxide is generated by the current produced by the battery (usually at the anode). Perforation has occurred as rapidly as 6 hours after ingestion.

Homemade Cells

Almost any liquid or moist object that has enough ions to be electrically conductive can serve as the electrolyte for a cell. As a novelty or science demonstration, it is possible to insert two electrodes made of different metals into a lemon, potato, etc. and generate small amounts of electricity. "Two-potato clocks" are also widely available in hobby and toy stores; they consist of a pair of cells, each consisting of a potato (lemon, et cetera) with two electrodes inserted into it, wired in series to form a battery with enough voltage to power a digital clock. Homemade cells of this kind are of no practical use.

A voltaic pile can be made from two coins (such as a nickel and a penny) and a piece of paper towel dipped in salt water. Such a pile generates a very low voltage but, when many are stacked in series, they can replace normal batteries for a short time.

Sony has developed a biological battery that generates electricity from sugar in a way that is similar to the processes observed in living organisms. The battery generates electricity through the use of enzymes that break down carbohydrates.

Lead acid cells can easily be manufactured at home, but a tedious charge/discharge cycle is needed to 'form' the plates. This is a process in which lead sulfate forms on the plates, and during charge is converted to lead dioxide (positive plate) and pure lead (negative plate). Repeating this process results in a microscopically rough surface, increasing the surface area. This increases the current the cell can deliver. Aluminium–air batteries can be produced with high-purity aluminium. Aluminium foil batteries will produce some electricity, but are not efficient, in part because a significant amount of (combustible) hydrogen gas is produced.

Smartdust

Smartdust is a system of many tiny microelectromechanical systems (MEMS) such as sensors, robots, or other devices, that can detect, for example, light, temperature, vibration, magnetism, or chemicals. They are usually operated on a computer network wirelessly and are distributed over some area to perform tasks, usually sensing through radio-frequency identification. Without an antenna of much greater size the range of tiny smart dust communication devices is measured in a few millimetres and they may be vulnerable to electromagnetic disablement and destruction by microwave exposure.

Design and Engineering

The concepts for Smart Dust emerged from a workshop at RAND in 1992 and a series of DARPA ISAT studies in the mid-1990s due to the potential military applications of the technology. The work was strongly influenced by work at UCLA and the University of Michigan during that period, as well as science fiction authors Stanislaw Lem, Neal Stephenson and Vernor Vinge. The first public presentation of the concept by that name was at the American Vacuum Society meeting in Anaheim in 1996.

A Smart Dust research proposal was presented to DARPA written by Kristofer S. J. Pister, Joe Kahn, and Bernhard Boser, all from the University of California, Berkeley, in 1997. The proposal, to build wireless sensor nodes with a volume of one cubic millimetre, was selected for funding in 1998. The project led to a working mote smaller than a grain of rice, and larger "COTS Dust" devices kicked off the TinyOS effort at Berkeley.

The concept was later expanded upon by Kris Pister in 2001. A recent review discusses various techniques to take smartdust in sensor networks beyond millimetre dimensions to the micrometre level.

The Ultra-Fast Systems component of the Nanoelectronics Research Centre at the University of Glasgow is a founding member of a large international consortium which is developing a related concept: smart specks.

Radioisotope Thermoelectric Generator

A radioisotope thermoelectric generator (RTG, RITEG) is an electrical generator that uses an array of thermocouples to convert the heat released by the decay of a suitable radioactive material into electricity by the Seebeck effect.

RTGs have been used as power sources in satellites, space probes and unmanned remote facilities such as a series of lighthouses built by the former Soviet Union inside the Arctic Circle. RTGs are usually the most desirable power source for robotic or unmaintained situations that need a few hundred watts (or less) of power for durations too long for fuel cells, batteries, or generators to provide economically and in places where solar cells are not practical. Safe use of RTGs requires containment of the radioisotopes long after the productive life of the unit.

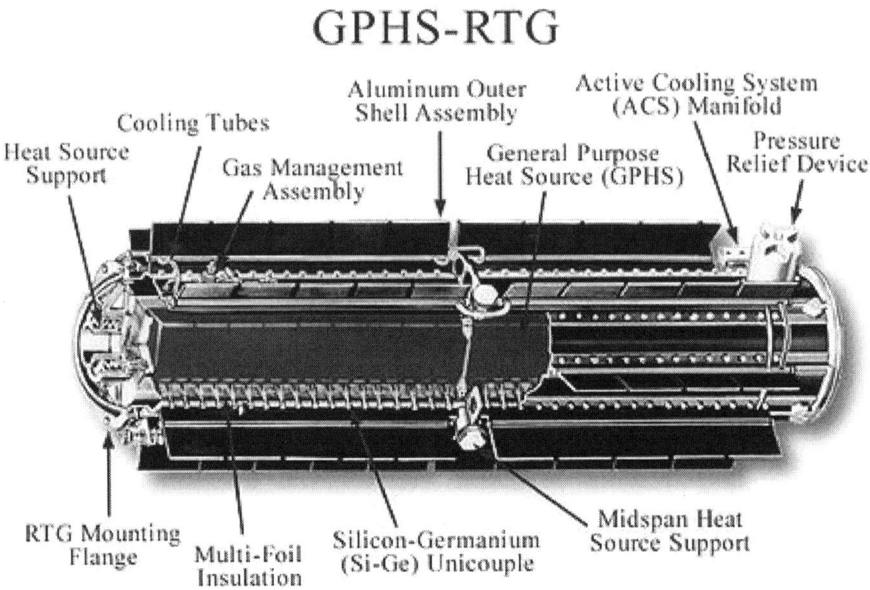

Figure: *Diagram of an RTG used on the Cassini probe*

In the same brief letter where he introduced the communications satellite, Arthur C. Clarke suggested that, with respect to spacecraft, "the operating period might be indefinitely prolonged by the use of thermocouples."

RTGs were developed in the US during the late 1950s by Mound Laboratories in Miamisburg, Ohio under contract with the United States Atomic Energy Commission. The project was led by Dr. Bertram C. Blanke.

The first RTG launched into space by the United States was SNAP 3 in 1961, aboard the Navy Transit 4A spacecraft. One of the first terrestrial uses of RTGs was in 1966 by the US Navy at uninhabited Fairway Rock in Alaska. RTGs were used at that site until 1995.

A common RTG application is spacecraft power supply. Systems for Nuclear Auxiliary Power (SNAP) units were used for probes that travelled far from the Sun rendering solar panels impractical. As such, they were used with Pioneer 10, Pioneer 11, Voyager 1, Voyager 2, Galileo, Ulysses, Cassini, New Horizons and the Mars Science Laboratory. RTGs were used to power the two Viking landers and for the scientific experiments left on the Moon by the crews of Apollo 12 through 17 (SNAP 27s). Because the Apollo 13 moon landing was aborted, its RTG rests in the South Pacific ocean, in the vicinity of the Tonga Trench. RTGs were also used for the Nimbus, Transit and LES satellites. By comparison, only a few space vehicles have been launched using full-fledged nuclear reactors: the Soviet RORSAT series and the American SNAP-10A.

In addition to spacecraft, the Soviet Union constructed many unmanned lighthouses and navigation beacons powered by RTGs. Powered by strontium-90 (^{90}Sr), they are very reliable and provide a steady source of power. Critics argue that they could cause environmental and security problems as leakage or theft of the radioactive material could pass unnoticed for years, particularly as the locations of some of these lighthouses are no longer known due to poor record keeping. In one instance, the radioactive compartments were opened by a thief. In another case, three woodsmen in Georgia came across two ceramic RTG heat sources that had been stripped of their shielding. Two of the three were later hospitalized with severe radiation burns after carrying the sources on their backs. The units were eventually recovered and isolated.

There are approximately 1,000 such RTGs in Russia. All of them have long exhausted their 10-year engineered life spans. They are likely no longer functional, and may be in need of dismantling. Some of them have become the prey of metal hunters, who strip the RTGs' metal casings, regardless of the risk of radioactive contamination.

The United States Air Force uses RTGs to power remote sensing stations for *Top-ROCC* and *Save-Igloo* radar systems predominantly located in Alaska.

In the past, small "plutonium cells" (very small ^{238}Pu-powered RTGs) were used in implanted heart pacemakers to ensure a very long "battery life". As of 2004, about 90 were still in use.

Design

The design of an RTG is simple by the standards of nuclear technology: the main component is a sturdy container of a radioactive material (the fuel). Thermocouples are placed in the walls of the container, with the outer end of each thermocouple connected to a heat sink. Radioactive decay of the fuel produces heat which flows through the thermocouples to the heat sink, generating electricity in the process.

A thermocouple is a thermoelectric device that converts thermal energy directly into electrical energy using the Seebeck effect. It is made of two kinds of metal (or semiconductors) that can both conduct electricity. They are connected to each other in a closed loop. If the two junctions are at different temperatures, an electric current will flow in the loop.

Criteria

The radioactive material used in RTGs must have several characteristics:

1. It should produce high energy radiation. Energy release per decay is proportional to power production per mole. Alpha decays in general release about 10 times as much energy as the beta decay of strontium-90 or cesium-137.
2. Radiation must be of a type easily absorbed and transformed into thermal radiation, preferably alpha radiation. Beta radiation can emit considerable gamma/X-ray radiation through bremsstrahlung secondary radiation production and therefore requires heavy shielding. Isotopes must not produce significant amounts of gamma, neutron radiation or penetrating radiation in general through other decay modes or decay chain products.
3. Its half-life must be long enough so that it will release energy at a relatively constant rate for a reasonable amount of time. The amount of energy released per time (power) of a given quantity is inversely proportional to half-life. An isotope with twice the half-life and the same energy per decay will release power at half the rate per mole. Typical half-lives for radioisotopes used in RTGs are therefore several decades, although isotopes with shorter half-lives could be used for specialised applications.
4. For spaceflight use, the fuel must produce a large amount of power per mass and volume (density). Density and weight are not as important for terrestrial use, unless there are size restrictions. The decay energy can be calculated if the energy of radioactive radiation or the mass loss before and after radioactive decay is known.

Selection of Isotopes

The first two criteria limit the number of possible fuels to fewer than 30 atomic isotopes within the entire table of nuclides. Plutonium-238, curium-244 and strontium-90 are the most often cited candidate isotopes, but other isotopes such as polonium-210, promethium-147, caesium-137, cerium-144, ruthenium-106, cobalt-60, curium-242, americium-241 and thulium isotopes have also been studied.

^{238}Pu, ^{90}Sr

Plutonium-238 has the lowest shielding requirements and longest half-life; its power output is 0.54 kilowatts per kilogram. Only three candidate isotopes meet the last criterion (not all are listed above) and need less than 25 mm of lead shielding to block the radiation. ^{238}Pu (the best of these three) needs less than 2.5 mm, and in many cases no shielding is needed in a ^{238}Pu RTG, as the casing itself is adequate. ^{238}Pu has become the most widely used fuel for RTGs, in the form of plutonium(IV) oxide (PuO_2). ^{238}Pu has a half-life of 87.7 years, reasonable power density, and exceptionally low gamma and neutron radiation levels.

Strontium-90 also requires little shielding, as it decays by β emission, with negligible γ emission. While its half life of 28.8 years is much shorter than that of ^{238}Pu, it also has a much lower decay energy. Thus its power density is only 0.46 kilowatts per kilogram. Because the energy output is lower it reaches lower temperatures than ^{238}Pu, which results in lower RTG efficiency. ^{90}Sr is a high yield waste product of nuclear fission and is available in large quantities at a low price.

^{210}Po

Some prototype RTGs, first built in 1958 by the US Atomic Energy Commission, have used polonium-210. This isotope provides phenomenal power density because of its high radioactive activity, but has limited use because of its very short half-life of 138 days. A kilogram of pure ^{210}Po in the form of a cube would be about 48 mm (about 2 inches) on a side and emit about 140 kW.

^{242}Cm, ^{244}Cm, ^{241}Am

Curium-242 and curium-244 have also been studied as well, but require heavy shielding for gamma and neutron radiation produced from spontaneous fission.

Americium-241 is a potential candidate isotope with a longer half-life than ^{238}Pu: ^{241}Am has a half-life of 432 years and could hypothetically power a device for centuries. However, the power density of ^{241}Am is only 1/4 that of ^{238}Pu, and ^{241}Am produces more penetrating radiation through decay chain products than ^{238}Pu and needs about 18 mm of lead shielding. Even so, its shielding requirements in an RTG are the second lowest of all possible isotopes: only ^{238}Pu requires less. With a current global shortage of ^{238}Pu, a closer look is being given to ^{241}Am.

Life Span

Most RTGs use ^{238}Pu, which decays with a half-life of 87.7 years. RTGs using this material will therefore diminish in power output by $1"0.5^{1/87.74} = 0.787\%$ of their capacity per year. 23 years after production, such an RTG will have decreased in power by 16.6%, i.e. providing 83.4% of its initial output. Thus, with a starting capacity of 470 W, after 23 years it would have a capacity of 392 W.

***Figure:** ^{90}Sr-powered Soviet RTGs in dilapidated and vandalized condition.*

However, the bi-metallic thermocouples used to convert thermal energy into electrical energy degrade as well; at the beginning of 2001, the power generated by the Voyager RTGs had dropped to 315 W for Voyager 1 and to 319 W for Voyager 2. Therefore in early 2001, the RTGs were working at about 67% of their original capacity instead of the expected 83.4%.

This life span was of particular importance during the Galileo mission. Originally intended to launch in 1986, it was delayed by the Space Shuttle Challenger accident. Because of this unforeseen event, the probe had to sit in storage for 4 years before launching in 1989. Subsequently, its RTGs had decayed somewhat, necessitating replanning the power budget for the mission.

Efficiency

RTGs use thermoelectric couples or "thermocouples" to convert heat from the radioactive material into electricity. Thermocouples, though very reliable and long-lasting, are very inefficient; efficiencies above 10% have never been achieved and most RTGs have efficiencies between 3–7%. Thermoelectric materials in space missions to date have included silicon–germanium alloys, lead telluride and tellurides of antimony, germanium and silver (TAGS). Studies have been done on improving efficiency by using other technologies to generate electricity from heat. Achieving higher efficiency would mean less radioactive fuel is needed to produce the same amount of power, and therefore a lighter overall weight for the generator. This is a critically important factor in spaceflight launch cost considerations.

A thermionic converter—an energy conversion device which relies on the principle of thermionic emission—can achieve efficiencies between

10–20%, but requires higher temperatures than those at which standard RTGs run. Some prototype ^{210}Po RTGs have used thermionics, and potentially other extremely radioactive isotopes could also provide power by this means, but short half-lives make these unfeasible. Several space-bound nuclear reactors have used thermionics, but nuclear reactors are usually too heavy to use on most space probes.

Thermophotovoltaic cells work by the same principles as a photovoltaic cell, except that they convert infrared light emitted by a hot surface rather than visible light into electricity. Thermophotovoltaic cells have an efficiency slightly higher than thermocouples and can be overlaid on top of thermocouples, potentially doubling efficiency. Systems with radioisotope generators simulated by electric heaters have demonstrated efficiencies of 20%, but have not been tested with actual radioisotopes. Some theoretical thermophotovoltaic cell designs have efficiencies up to 30%, but these have yet to be built or confirmed. Thermophotovoltaic cells and silicon thermocouples degrade faster than thermocouples, especially in the presence of ionizing radiation.

Dynamic generators can provide power at more than 4 times the conversion efficiency of RTGs. NASA and DOE have been developing a next-generation radioisotope-fuelled power source called the Stirling Radioisotope Generator (SRG) that uses free-piston Stirling engines coupled to linear alternators to convert heat to electricity. SRG prototypes demonstrated an average efficiency of 23%. Greater efficiency can be achieved by increasing the temperature ratio between the hot and cold ends of the generator. The use of non-contacting moving parts, non-degrading flexural bearings, and a lubrication-free and hermetically sealed environment have, in test units, demonstrated no appreciable degradation over years of operation. Experimental results demonstrate that an SRG could continue running for decades without maintenance. Vibration can be eliminated as a concern by implementation of dynamic balancing or use of dual-opposed piston movement. Potential applications of a Stirling radioisotope power system include exploration and science missions to deep-space, Mars, and the Moon.

The increased efficiency of the SRG may be demonstrated by a theoretical comparison of thermodynamic properties, as follows. These calculations are simplified and do not account for the decay of thermal power input due to the long half-life of the radioisotopes used in these generators. The assumptions for this analysis include that both systems are operating at steady state under the conditions observed in experimental procedures. Both generators can be simplified to heat engines to be able to compare their current efficiencies to their corresponding Carnot efficiencies. The system is assumed to be the components, apart from the heat source and heat sink.,

The thermal efficiency, denoted η_{th}, is given by:

$$\eta_{th} = \frac{\text{Desired Output}}{\text{Required Input}} = \frac{W'_{out}}{Q'_{in}}$$

Where primes (') denote the time derivative.

From a general form of the First Law of Thermodynamics, in rate form:

$$\Delta E'^{sys} = Q'_{in} + W'_{in} - Q'_{out} - W'_{out}$$

Assuming the system is operating at steady state and $W'_{in} = 0$,

$$W'_{out} = Q'_{in} - Q'_{out}$$

η_{th}, then, can be calculated to be 110 W / 2000 W = 5.5% (or 140 W / 500 W = 28% for the SRG). Additionally, the Second Law efficiency, denoted η_{II}, is given by:

$$\eta_{II} = \frac{\eta_{th}}{\eta_{th,rev}}$$

Where $\eta_{th,rev}$ is the Carnot efficiency, given by:

$$\eta_{th} = 1 - \frac{T_{heatsink}}{T_{heatsource}}$$

In which $T_{heat\ sink}$ is the external temperature (which has been measured to be 510 K for the MMRTG (Multi-Mission RTG) and 363 K for the SRG) and $T_{heat\ source}$ is the temperature of the MMRTG, assumed 823 K (1123 K for the SRG). This yields a Second Law efficiency of 14.46% for the MMRTG (or 41.37% for the SRG).

Safety

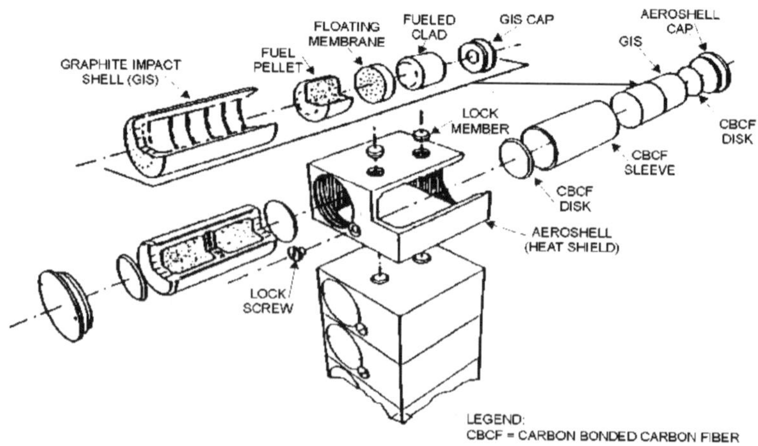

***Figure:** Diagram of a stack of general purpose heat source modules as used in RTGs*

Radioactive Contamination

RTGs pose a risk of radioactive contamination: if the container holding the fuel leaks, the radioactive material may contaminate the environment. For spacecraft, the main concern is that if an accident were to occur during launch or a subsequent passage of a spacecraft close to Earth, harmful material could be released into the atmosphere; therefore their use in spacecraft and elsewhere has attracted controversy.

However, this event is not considered likely with current RTG cask designs. For instance, the environmental impact study for the Cassini–Huygens probe launched in 1997 estimated the probability of contamination accidents at various stages in the mission. The probability of an accident occurring which caused radioactive release from one or more of its 3 RTGs (or from its 129 radioisotope heater units) during the first 3.5 minutes following launch was estimated at 1 in 1,400; the chances of a release later in the ascent into orbit were 1 in 476; after that the likelihood of an accidental release fell off sharply to less than 1 in a million. If an accident which had the potential to cause contamination occurred during the launch phases (such as the spacecraft failing to reach orbit), the probability of contamination actually being caused by the RTGs was estimated at about 1 in 10. In any event, the launch was successful and Cassini–Huygens reached Saturn.

The plutonium-238 used in these RTGs has a half-life of 87.74 years, in contrast to the 24,110 year half-life of plutonium-239 used in nuclear weapons and reactors. A consequence of the shorter half-life is that plutonium-238 is about 275 times more radioactive than plutonium-239 (i.e. 17.3 curies (640 GBq)/g compared to 0.063 curies (2.3 GBq)/g). For instance, 3.6 kg of plutonium-238 undergoes the same number of radioactive decays per second as 1 tonne of plutonium-239. Since the morbidity of the two isotopes in terms of absorbed radioactivity is almost exactly the same, plutonium-238 is around 275 times more toxic by weight than plutonium-239.

The alpha radiation emitted by either isotope will not penetrate the skin, but it can irradiate internal organs if plutonium is inhaled or ingested. Particularly at risk is the skeleton, the surface of which is likely to absorb the isotope, and the liver, where the isotope will collect and become concentrated.

There have been several known accidents involving RTG-powered spacecraft:

1. The first one was a launch failure on 21 April 1964 in which the U.S. Transit-5BN-3 navigation satellite failed to achieve

orbit and burnt up on re-entry north of Madagascar. The 17,000 Ci (630 TBq) plutonium metal fuel in its SNAP-9a RTG was injected into the atmosphere over the Southern Hemisphere where it burnt up, and traces of plutonium-238 were detected in the area a few months later.

2. The second was the Nimbus B-1 weather satellite whose launch vehicle was deliberately destroyed shortly after launch on 21 May 1968 because of erratic trajectory. Launched from the Vandenberg Air Force Base, its SNAP-19 RTG containing relatively inert plutonium dioxide was recovered intact from the seabed in the Santa Barbara Channel five months later and no environmental contamination was detected.
3. In 1969 the launch of the first Lunokhod lunar rover mission failed, spreading polonium 210 over a large area of Russia
4. The failure of the Apollo 13 mission in April 1970 meant that the Lunar Module reentered the atmosphere carrying an RTG and burnt up over Fiji. It carried a SNAP-27 RTG containing 44,500 Ci (1,650 TBq) of plutonium dioxide which survived reentry into the Earth's atmosphere intact, as it was designed to do, the trajectory being arranged so that it would plunge into 6–9 kilometres of water in the Tonga trench in the Pacific Ocean. The absence of plutonium-238 contamination in atmospheric and seawater sampling confirmed the assumption that the cask is intact on the seabed. The cask is expected to contain the fuel for at least 10 half-lives (i.e. 870 years). The US Department of Energy has conducted seawater tests and determined that the graphite casing, which was designed to withstand reentry, is stable and no release of plutonium should occur. Subsequent investigations have found no increase in the natural background radiation in the area. The Apollo 13 accident represents an extreme scenario because of the high re-entry velocities of the craft returning from cis-lunar space (the region between Earth's atmosphere and the Moon). This accident has served to validate the design of later-generation RTGs as highly safe.

To minimize the risk of the radioactive material being released, the fuel is stored in individual modular units with their own heat shielding. They are surrounded by a layer of iridium metal and encased in high-strength graphite blocks. These two materials are corrosion- and heat-resistant. Surrounding the graphite blocks is an aeroshell, designed to protect the entire assembly against the heat of reentering the Earth's atmosphere. The plutonium fuel is also stored in a ceramic form that is heat-resistant, minimising the risk of vaporization and aerosolization. The ceramic is also highly insoluble.

The most recent accident involving a spacecraft RTG was the failure of the Russian Mars 96 probe launch on 16 November 1996. The two RTGs onboard carried in total 200 g of plutonium and are assumed to have survived reentry as they were designed to do. They are thought to now lie somewhere in a northeast-southwest running oval 320 km long by 80 km wide which is centred 32 km east of Iquique, Chile.

Figure: *A SNAP-27 RTG deployed by the astronauts of Apollo 14 identical to the one lost in the reentry of Apollo 13*

Many Beta-M RTGs produced by the Soviet Union to power lighthouses and beacons have become orphaned sources of radiation. Several of these units have been illegally dismantled for scrap metal resulting in the complete exposure of the Sr-90 source, fallen into the ocean, or have defective shielding due to poor design or physical damage. The US Department of Defence cooperative threat reduction program has expressed concern that material from the Beta-M RTGs can be used by terrorists to construct a dirty bomb.

28 U.S. space missions have safely flown radioisotope energy sources since 1961.

Nuclear Fission

RTGs and nuclear power reactors use very different nuclear reactions. Nuclear power reactors use controlled nuclear fission. When an atom of U-235 or Pu-239 fuel fissions, neutrons are released that trigger additional fissions in a chain reaction at a rate that can be controlled with neutron absorbers. This is an advantage in that power can be varied with demand or shut off entirely for maintenance. It is also a disadvantage in that care is needed to avoid uncontrolled operation at dangerously high power levels.

Chain reactions do not occur in RTGs, so heat is produced at an unchangeable, though steadily decreasing rate that depends only on the amount of fuel isotope and its half-life. An accidental power excursion is impossible. However, if a launch or re-entry accident occurs and the fuel is dispersed, the combined power output of the now radionuclides set free does not drop. In an RTG, heat generation cannot be varied with demand or shut off when not needed. Therefore, auxiliary power supplies (such as rechargeable batteries) may be needed to meet peak demand, and adequate cooling must be provided at all times including the pre-launch and early flight phases of a space mission.

RTG for Interstellar Probes

RTG have been proposed for use on realistic interstellar precursor missions and interstellar probes. An example of this is the Innovative Interstellar Explorer (2003–current) proposal from NASA. A RTG using ^{241}Am was proposed for this type of mission in 2002. This could support mission extensions up to 1000 years on the interstellar probe, because the power output would be more stable in the long-term than plutonium. Other isotopes for RTG were also examined in the study, looking at traits such as watt/gram, half-life, and decay products. An interstellar probe proposal from 1999 suggested using three advanced radioisotope power source (ARPS).

The RTG electricity can be used for powering scientific instruments and communication to Earth on the probes. One mission proposed using the electricity to power ion engines, calling this method radioisotope electric propulsion (REP).

5

Graphene for Energy Production and Storage

Graphene is pure carbon in the form of a very thin, nearly transparent sheet, basically one atom thick. It is remarkably strong for its very low weight (100 times stronger than steel) and it conducts heat and electricity with great efficiency. While scientists had theorized about graphene for decades, it was first produced in the lab in 2004. Because it is virtually two-dimensional, it interacts oddly with light and with other materials. Researchers have identified the bipolar transistor effect, ballistic transport of charges and large quantum oscillations.

Technically, graphene is a crystalline allotrope of carbon with 2-dimensional properties. In graphene, carbon atoms are densely packed in a regular sp^2-bonded atomic-scale chicken wire (hexagonal) pattern. Graphene can be described as a one-atom thick layer of graphite. It is the basic structural element of other allotropes, including graphite, charcoal, carbon nanotubes and fullerenes. It can also be considered as an indefinitely large aromatic molecule, the limiting case of the family of flat polycyclic aromatic hydrocarbons.

Graphene research has expanded quickly since the substance was first isolated in 2004. Research was informed by theoretical descriptions of graphene's composition, structure and properties, which had all been calculated decades earlier. High-quality graphene also proved to be surprisingly easy to isolate, making more research possible. Andre

Geim and Konstantin Novoselov at the University of Manchester won the Nobel Prize in Physics in 2010 "for groundbreaking experiments regarding the two-dimensional material graphene".

"Graphene" is a combination of graphite and the suffix -ene, named by Hanns-Peter Boehm, who described single-layer carbon foils in 1962.

The term *graphene* first appeared in 1987 to describe single sheets of graphite as a constituent of graphite intercalation compounds (GICs); conceptually a GIC is a crystalline salt of the intercalant and graphene. The term was also used in early descriptions of carbon nanotubes, as well as for epitaxial graphene and polycyclic aromatic hydrocarbons.

The IUPAC compendium of technology states: "previously, descriptions such as graphite layers, carbon layers, or carbon sheets have been used for the term graphene... it is incorrect to use for a single layer a term which includes the term graphite, which would imply a three-dimensional structure. The term graphene should be used only when the reactions, structural relations or other properties of individual layers are discussed."

Geim defined "isolated or free-standing graphene" as "graphene is a single atomic plane of graphite, which – and this is essential – is sufficiently isolated from its environment to be considered free-standing." This definition is narrower than the IUPAC definition and refers to cleaved, transferred and suspended graphene. Other forms of graphene, such as graphene grown on various metals, can become free-standing if, for example, suspended or transferred to silicon dioxide (SiO2) or silicon carbide (after its passivation with hydrogen).

Graphene can be considered an "infinite alternant" (only six-member carbon ring) polycyclic aromatic hydrocarbon (PAH).

History

In 1859 Benjamin Collins Brodie was aware of the highly lamellar structure of thermally reduced graphite oxide.

The structure of graphite was solved in 1916. by the related method of powder diffraction, It was studied in detail by V. Kohlschütter and P. Haenni in 1918, who also described the properties of graphite oxide paper. Its structure was determined from single-crystal diffraction in 1924.

The theory of graphene was first explored by P. R. Wallace in 1947 as a starting point for understanding the electronic properties of 3D graphite. The emergent massless Dirac equation was first pointed out by Gordon Walter Semenoff and David P. DeVincenzo and Eugene J.

Mele. Semenoff emphasized the occurrence in a magnetic field of an electronic Landau level precisely at the Dirac point. This level is responsible for the anomalous integer quantum Hall effect.

The earliest TEM images of few-layer graphite were published by G. Ruess and F. Vogt in 1948. Later, single graphene layers were also observed directly by electron microscopy. Before 2004 intercalated graphite compounds were studied under a transmission electron microscope (TEM). Researchers occasionally observed thin graphitic flakes ("few-layer graphene") and possibly even individual layers. An early, detailed study on few-layer graphite dates to 1962.

Starting in the 1970s single layers of graphite were grown epitaxially on top of other materials. This "epitaxial graphene" consists of a single-atom-thick hexagonal lattice of sp^2-bonded carbon atoms, as in free-standing graphene. However, there is significant charge transfer from the substrate to the epitaxial graphene, and, in some cases, hybridization between the d-orbitals of the substrate atoms and π orbitals of graphene, which significantly alters the electronic structure of epitaxial graphene.

Single layers of graphite were also observed by transmission electron microscopy within bulk materials, in particular inside soot obtained by chemical exfoliation. Efforts to make thin films of graphite by mechanical exfoliation started in 1990, but nothing thinner than 50 to 100 layers was produced before 2004.

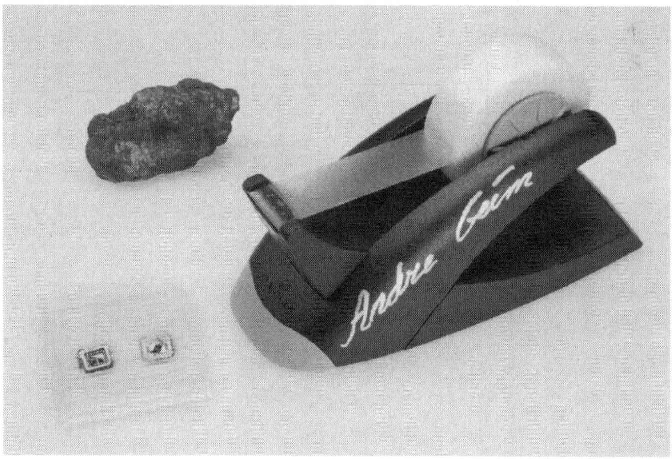

***Figure:** A lump of graphite, a graphene transistor and a tape dispenser. Donated to the Nobel Museum in Stockholm by Andre Geim and Konstantin Novoselov in 2010.*

Initial attempts to make atomically thin graphitic films employed exfoliation techniques similar to the drawing method. Multilayer samples down to 10 nm in thickness were obtained. Old papers were unearthed in which researchers tried to isolate graphene starting with intercalated compounds. These papers reported the observation of very thin graphitic fragments (possibly monolayers) by transmission electron microscopy. Neither of the earlier observations was sufficient to "spark the graphene gold rush", which awaited macroscopic samples of extracted atomic planes.

One of the very first patents pertaining to the production of graphene was filed in October, 2002 (US Pat. 7071258). Entitled, "Nano-scaled Graphene Plates", this patent detailed one of the very first large scale graphene production processes. Two years later, in 2004 Andre Geim and Kostya Novoselov at University of Manchester extracted single-atom-thick crystallites from bulk graphite. They pulled graphene layers from graphite and transferred them onto thin SiO2 on a silicon wafer in a process called either micromechanical cleavage or the Scotch tape technique. The SiO2 electrically isolated the graphene and weakly interacted with it, providing nearly charge-neutral graphene layers. The silicon beneath the SiO2 could be used as a "back gate" electrode to vary the charge density in the graphene over a wide range. They may not have been the first to use this technique— US 6667100, filed in 2002, describes how to process commercially available flexible expanded graphite to achieve a graphite thickness of 0.01 thousandth of an inch. The key to success was high-throughput visual recognition of graphene on a properly chosen substrate, which provides a small but noticeable optical contrast.

The cleavage technique led directly to the first observation of the anomalous quantum Hall effect in graphene, which provided direct evidence of graphene's theoretically predicted Berry's phase of massless Dirac fermions. The effect was reported soon after by Philip Kim and Yuanbo Zhang in 2005. These experiments started after the researchers observed colleagues who were looking for the quantum Hall effect and Dirac fermions in bulk graphite.

Even though graphene on nickel and on silicon carbide have both existed in the laboratory for decades, graphene mechanically exfoliated on SiO2 provided the first proof of the Dirac fermion nature of electrons.

Properties

Structure

The atomic structure of isolated, single-layer graphene was studied by transmission electron microscopy (TEM) on sheets of graphene suspended between bars of a metallic grid. Electron diffraction patterns showed the expected honeycomb lattice. Suspended graphene also showed "rippling" of the flat sheet, with amplitude of about one nanometre.

These ripples may be intrinsic to the material as a result of the instability of two-dimensional crystals, or may originate from the ubiquitous dirt seen in all TEM images of graphene. Atomic resolution real-space images of isolated, single-layer graphene on SiO_2 substrates are available via scanning tunneling microscopy. Photoresist residue, which must be removed to obtain atomic-resolution images, may be the "adsorbates" observed in TEM images, and may explain the observed rippling. Rippling on SiO_2 is caused by conformation of graphene to the underlying SiO_2, and is not intrinsic.

Graphene sheets in solid form usually show evidence in diffraction for graphite's (002) layering. This is true of some single-walled nanostructures. However, unlayered graphene with only (hk0) rings has been found in the core of presolar graphite onions. TEM studies show faceting at defects in flat graphene sheets and suggest a role for two-dimensional crystallization from a melt.

Graphene can self-repair holes in its sheets, when exposed to molecules containing carbon, such as hydrocarbons. Bombarded with pure carbon atoms, the atoms perfectly align into hexagons, completely filling the holes.

Chemical

Graphene is the only form of carbon (and generally all solid materials) in which each single atom is in exposure for chemical reaction from two sides (due to the 2D structure). It is known that carbon atoms at the edge of graphene sheets have special chemical reactivity, and graphene has the highest ratio of edgy carbons (in comparison with similar materials such as carbon nanotubes). In addition, various types of defects within the sheet, which are very common, increase the chemical reactivity. The onset temperature of reaction between the basal plane of single-layer graphene and oxygen gas is below 260 °C and the graphene burns at very low temperature (e.g., 350 °C). In fact, graphene is chemically the most reactive form of carbon, owing to the lateral availability of carbon atoms. Graphene is commonly modified with oxygen- and nitrogen-containing functional groups and analyzed by infrared spectroscopy and X-ray photoelectron spectroscopy. However, determination of structures of graphene with oxygen- and nitrogen-containing functional groups is a difficult task unless the structures are well controlled.

In 2013, Stanford University physicists reported that sheets of Graphene one atom thick are a hundred times more chemically reactive than thicker sheets.

Physical

The carbon–carbon bond length in graphene is about 0.142 nanometres. Graphene sheets stack to form graphite with an interplanar spacing of 0.335 nm.

Electronic

Graphene is a semi-metal or zero-gap semiconductor. Four electronic properties separate it from other condensed matter systems.

Electronic Spectrum

Electrons propagating through graphene's honeycomb lattice effectively lose their mass, producing quasi-particles that are described by a 2D analogue of the Dirac equation rather than the Schrödinger equation for spin-1/2 particles.

Dispersion Relation

Using a conventional tight-binding model the dispersion relation produces energy of the electrons with wave vector k is

$$E = \pm\sqrt{\gamma_0^2\left(1 + 4\cos^2\frac{k_y a}{2} + 4\cos\frac{k_y a}{2}\cdot\cos\frac{k_x\sqrt{3}a}{2}\right)}$$

with the nearest-neighbour hopping energy $\gamma_0 \approx 2.8$ eV and the lattice constant a H" 2.46 E. The conduction and valence bands, respectively, correspond to the different signs; they touch each other at six points, the "K-values" of the two-dimensional hexagonal Brillouin zone. Two of these six points are independent, while the rest are equivalent by symmetry. In the vicinity of the K-points the energy depends *linearly* on the wave vector, similar to a relativistic particle. Since an elementary cell of the lattice has a basis of two atoms, the wave function has an effective 2-spinor structure.

As a consequence, at low energies, even neglecting the true spin, the electrons can be described by an equation that is formally equivalent to the massless Dirac equation. Hence, the electrons and holes are called Dirac fermions and the six corners are called the Dirac points. This pseudo-relativistic description is restricted to the chiral limit, i.e., to vanishing rest mass M_0, which leads to interesting additional features:

$$v_F\,\vec{\sigma}\cdot\nabla\psi(\mathrm{r}) = E\psi(\mathrm{r}).$$

Here $v_F \sim 10^6$ m/s (.003 c) is the Fermi velocity in graphene, which replaces the velocity of light in the Dirac theory; $\vec{\sigma}$ is the vector of the Pauli matrices, $\psi(\mathrm{r})$ is the two-component wave function of the electrons, and E is their energy.

The equation describing the electrons' linear dispersion relation is

$$E = \hbar v_F \sqrt{k_x^2 + k_y^2} ;$$

where the wavevector k is measured from the Dirac points (the zero of energy is chosen here to coincide with the Dirac points). The equation uses a pseudospin matrix formula that describes two sublattices of the honeycomb lattice.

Single-Atom Wave Propagation

Electron waves in graphene propagate within a single-atom layer, making them sensitive to the proximity of other materials such as high-ê dielectrics, superconductors and ferromagnetics.

Conductance

Graphene electrons can cover micrometre distances without scattering, even at room temperature.

Electrical resistance in 40-nanometre-wide nanoribbons of epitaxially-created graphene changes in discrete steps following quantum mechanical principles. Electrons travel ballistically, similar to those observed in cylindrical carbon nanotubes. The ribbons' theoretical conductance exceeds predictions by a factor of 10. Electrons can move tens or hundreds of microns without scattering. The ribbons can act more like optical waveguides or quantum dots, allowing electrons to flow smoothly along the ribbons' edges. By contrast, in conductors such as copper, resistance increases in proportion to the length as electrons encounter impurities while moving through the conductor.

Transport is dominated by two modes. One is ballistic and temperature independent, while The other is thermally activated. Transport is protected from back-scattering, possibly reflecting ground-state properties of neutral graphene. At room temperature, the resistance of both modes is found to increase abruptly at a particular length— the ballistic mode at 16 micrometres and the other at 160 nanometres (1% of the former length).

The ribbons were grown on the edges of three-dimensional structures etched into silicon carbide wafers. When the wafers are heated to approximately 1,000 °C (1,830 °F), silicon is preferentially driven off along the edges, forming nanoribbons whose structure is determined by the pattern of the three-dimensional surface. The nanoribbons had perfectly smooth edges, annealed by the fabrication process. Electron mobility measurements surpassing one million correspond to a sheet resistance of one ohm per square— two orders of magnitude lower than in two-dimensional graphene.

Anomalous Quantum Hall Effect

The quantum Hall effect is a quantum mechanical version of the Hall effect, which is the production of transverse (perpendicular to the main current) conductivity in the presence of a magnetic field. The quantization of the Hall effect σ_{xy} at integer multiples (the "Landau level") of the basic quantity e^2 / h (where *e* is the elementary electric charge and *h* is Planck's constant) It can usually be observed only in very clean silicon or gallium arsenide solids at temperatures around 3 K and very high magnetic fields.

Graphene shows the quantum Hall effect with respect to conductivity-quantization: the effect is *anomalous* in that the sequence of steps is shifted by 1/2 with respect to the standard sequence and with an additional factor of 4. Graphene's Hall conductivity is $\sigma_{xy} = \pm 4 \cdot (N + 1/2) e^2 / h$, where *N* is the Landau level and the double valley and double spin degeneracies give the factor of 4. These anomalies are present at room temperature, i.e. at roughly 20 °C.

This behaviour is a direct result of graphene's massless Dirac electrons. In a magnetic field, their spectrum has a Landau level with energy precisely at the Dirac point. This level is a consequence of the Atiyah–Singer index theorem and is half-filled in neutral graphene, leading to the "+1/2" in the Hall conductivity. Bilayer graphene also shows the quantum Hall effect, but with only one of the two anomalies (i.e. $\sigma_{xy} = \pm 4 \cdot N \cdot e^2 / h$). In the second anomaly, the first plateau at *N=0* is absent, indicating that bilayer graphene stays metallic at the neutrality point.

Unlike normal metals, graphene's longitudinal esistance shows maxima rather than minima for integral values of the Landau filling factor in measurements of the Shubnikov–De Haas oscillations, whereby the term *integral* quantum Hall effect. These oscillations a how a phase shift of π, known as Berry's phase. Berry's phase arises due to the zero effective carrier mass near the Dirac points. The temperature dependence of the oscillations reveals that the carriers have a non-zero cyclotron mass, despite their zero effective mass.

Graphene samples prepared on nickel films, and on both the silicon face and carbon face of silicon carbide, show the anomalous effect directly in electrical measurements. Graphitic layers on the carbon face of silicon carbide show a clear Dirac spectrum in angle-resolved photoemission experiments, and the effect is observed in cyclotron resonance and tunneling experiments.

Strong Magnetic Fields

In magnetic fields above 10 Teslas or so additional plateaus of the Hall conductivity at $\sigma_{xy} = \nu e^2 / h$ with $\nu = 0, \pm 1, \pm 4$ are observed. A plateau at $\nu = 3$ and the fractional quantum Hall effect at $\nu = 1/3$ were also reported.

These observations with $\nu = 0, \pm 1, \pm 3, \pm 4$ indicate that the four-fold degeneracy (two valley and two spin degrees of freedom) of the Landau energy levels is partially or completely lifted.

Casimir Effect

The Casimir effect is an interaction between disjoint neutral bodies provoked by the fluctuations of the electrodynamical vacuum. Mathematically it can be explained by considering the normal modes of electromagnetic fields, which explicitly depend on the boundary (or matching) conditions on the interacting bodies' surfaces. Since graphene/ electromagnetic field interaction is strong for a one-atom-thick material, the Casimir effect is of growing interest.

van der Waals Force

The van der Waals force (or dispersion force) is also unusual, obeying an inverse cubic, asymptotic power law in contrast to the usual inverse quartic.

"Massive" Electrons

Graphene's unit cell has two identical carbon atoms and two zero-energy states: one in which the electron resides on atom A, the other in which the electron resides on atom B. Both states exist at exactly zero energy. However, if the two atoms in the unit cell are not identical, the situation changes. Hunt et al. show that placing hBN in contact with graphene can alter the potential felt at atom A versus atom B enough that the electrons develop a mass and accompanying band gap of about 30 meV.

The mass can be positive or negative. An arrangement that slightly raises the energy of an electron on atom A relative to atom B gives it a positive mass, while an arrangement that raises the energy of atom B produces a negative electron mass. The two versions behave alike and are indistinguishable via optical spectroscopy. An electron travelling from a positive-mass region to a negative-mass region must cross an intermediate region where its mass once again becomes zero. This region is gapless and therefore metallic. Metallic modes bounding semiconducting regions of opposite-sign mass is a hallmark of a topological phase and display much the same physics as topological insulators.

If the mass in graphene can be controlled, electrons can be confined to massless regions by surrounding them with massive regions, allowing the patterning of quantum dots, wires, and other mesoscopic structures. It also produces one-dimensional conductors along the boundary. These wires would be protected against backscattering and could carry currents without dissipation.

Electron Transport

Experimental results from transport measurements show that graphene has a remarkably high electron mobility at room temperature, with reported values in excess of 15,000 $cm^2 \cdot V^{-1} \cdot s^{-1}$. Additionally, the symmetry of the experimentally measured conductance indicates that hole and electron mobilities should be nearly the same. The mobility is nearly independent of temperature between 10 K and 100 K, which implies that the dominant scattering mechanism is defect scattering. Scattering by the acoustic phonons of graphene intrinsically limits room temperature mobility to 200,000 $cm^2 \cdot V^{-1} \cdot s^{-1}$ at a carrier density of 10^{12} cm^{-2}. The corresponding resistivity of the graphene sheet would be 10^{-6} $\Omega \cdot cm$. This is less than the resistivity of silver, the lowest known at room temperature. However, for graphene on SiO_2 substrates, scattering of electrons by optical phonons of the substrate is a larger effect at room temperature than scattering by graphene's own phonons. This limits mobility to 40,000 $cm^2 \cdot V^{-1} \cdot s^{-1}$.

Despite zero carrier density near the Dirac points, graphene exhibits a minimum conductivity on the order of $4e^2 / h$. The origin of this minimum conductivity is still unclear. However, rippling of the graphene sheet or ionized impurities in the SiO_2 substrate may lead to local puddles of carriers that allow conduction. Several theories suggest that the minimum conductivity should be $4e^2 / (\pi h)$; however, most measurements are of order $4e^2 / h$ or greater and depend on impurity concentration.

Graphene doped with various gaseous species (both acceptors and donors) can be returned to an undoped state by gentle heating in vacuum. Even for dopant concentrations in excess of 10^{12} cm^{-2} carrier mobility exhibits no observable change. Graphene doped with potassium in ultra-high vacuum at low temperature can reduce mobility 20-fold. The mobility reduction is reversible on heating the graphene to remove the potassium. Due to graphene's two dimensions, charge fractionalization (where the apparent charge of individual pseudoparticles in low-dimensional systems is less than a single quantum) is thought to occur. It may therefore be a suitable material for constructing quantum computers using anyonic circuits.

Optical

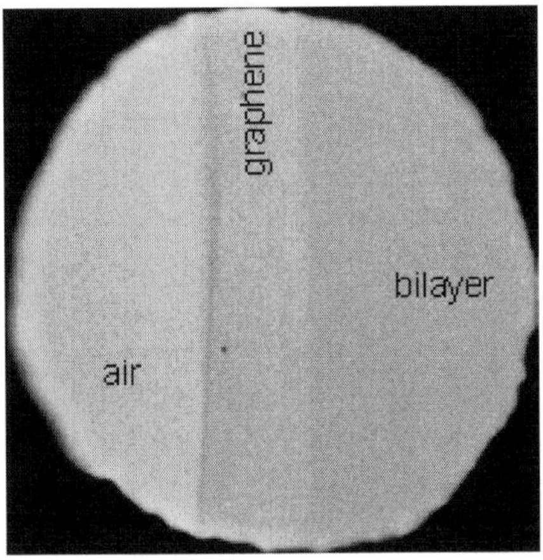

Figure: *Photograph of graphene in transmitted light. This one-atom-thick crystal can be seen with the naked eye because it absorbs approximately 2.3% of white light.*

Graphene's unique optical properties produce an unexpectedly high opacity for an atomic monolayer in vacuum, absorbing $\pi\alpha$ H" 2.3% of white light, where α is the fine-structure constant. This is a consequence of the "unusual low-energy electronic structure of monolayer graphene that features electron and hole conical bands meeting each other at the Dirac point... [which] is qualitatively different from more common quadratic massive bands". Based on the Slonczewski–Weiss–McClure (SWMcC) band model of graphite, the interatomic distance, hopping value and frequency cancel when optical conductance is calculated using Fresnel equations in the thin-film limit.

Although confirmed experimentally, the measurement is not precise enough to improve on other techniques for determining the fine-structure constant.

Graphene's band gap can be tuned from 0 to 0.25 eV (about 5 micrometre wavelength) by applying voltage to a dual-gate bilayer graphene field-effect transistor (FET) at room temperature. The optical response of graphene nanoribbons is tunable into the terahertz regime by an applied magnetic field. Graphene/graphene oxide systems exhibit electrochromic behaviour, allowing tuning of both linear and ultrafast optical properties.

A graphene-based Bragg grating (one-dimensional photonic crystal) has been fabricated and demonstrated its capability for excitation of surface electromagnetic waves in the periodic structure by using 633 nm He–Ne laser as the light source.

Saturable Absorption

Such unique absorption could become saturated when the input optical intensity is above a threshold value. This nonlinear optical behaviour is termed saturable absorption and the threshold value is called the saturation fluence. Graphene can be saturated readily under strong excitation over the visible to near-infrared region, due to the universal optical absorption and zero band gap. This has relevance for the mode locking of fibre lasers, where fullband mode locking has been achieved by graphene-based saturable absorber. Due to this special property, graphene has wide application in ultrafast photonics. Moreover, the optical response of graphene/graphene oxide layers can be tuned electrically. Saturable absorption in graphene could occur at the Microwave and Terahertz band, owing to its wideband optical absorption property. The microwave saturable absorption in graphene demonstrates the possibility of graphene microwave and terahertz photonics devices, such as microwave saturable absorber, modulator, polarizer, microwave signal processing and broad-band wireless access networks.

Nonlinear Kerr Effect

Under more intensive laser illumination, graphene could also possess a nonlinear phase shift due to the optical nonlinear Kerr effect. Based on a typical open and close aperture z-scan measurement, graphene possesses a giant non-linear Kerr coefficient of 10^{-7} $cm^2 \cdot W^{-1}$, almost nine orders of magnitude larger than that of bulk dielectrics. This suggests that graphene may be a nonlinear Kerr medium, paving the way for graphene-based nonlinear Kerr photonics such as a soliton.

Excitonic

First-principle calculations with quasiparticle corrections and many-body effects are performed to study the electronic and optical properties of graphene-based materials. The approach is described as three stages. With GW calculation, the properties of graphene-based materials are accurately investigated, including graphene, graphene nanoribbons, edge and surface functionalized armchair graphene nanoribbons, hydrogen saturated armchair graphene nanoribbons, Josephson effect in graphene SNS junctions with single localized defect and scaling properties in armchair graphene nanoribbons.

Thermal

Stability: Ab initio calculations show that a graphene sheet is thermodynamically unstable if its size is less than about 20 nm ("graphene is the least stable structure until about 6000 atoms") and becomes the most stable fullerene (as within graphite) only for molecules larger than 24,000 atoms.

Conductivity

The near-room temperature thermal conductivity of graphene was measured to be between (4.84±0.44) × 10^3 to (5.30±0.48) × 10^3 W·m^{-1}·K^{-1}. These measurements, made by a non-contact optical technique, are in excess of those measured for carbon nanotubes or diamonds. The isotopic composition, the ratio of ^{12}C to ^{13}C, has a significant impact on thermal conductivity, where isotopically pure ^{12}C graphene has higher conductivity than either a 50:50 isotope ratio or the naturally occurring 99:1 ratio. It can be shown by using the Wiedemann–Franz law, that the thermal conduction is phonon-dominated. However, for a gated graphene strip, an applied gate bias causing a Fermi energy shift much larger than k_BT can cause the electronic contribution to increase and dominate over the phonon contribution at low temperatures. The ballistic thermal conductance of graphene is isotropic.

Potential for this high conductivity can be seen by considering graphite, a 3D version of graphene that has basal plane thermal conductivity of over a 1000 W·m^{-1}·K^{-1} (comparable to diamond). In graphite, the c-axis (out of plane) thermal conductivity is over a factor of ~100 smaller due to the weak binding forces between basal planes as well as the larger lattice spacing. In addition, the ballistic thermal conductance of graphene is shown to give the lower limit of the ballistic thermal conductances, per unit circumference, length of carbon nanotubes.

Despite its 2-D nature, graphene has 3 acoustic phonon modes. The two in-plane modes (LA, TA) have a linear dispersion relation, whereas the out of plane mode (ZA) has a quadratic dispersion relation. Due to this, the T^2 dependent thermal conductivity contribution of the linear modes is dominated at low temperatures by the $T^{1.5}$ contribution of the out of plane mode. Some graphene phonon bands display negative Grüneisen parameters. At low temperatures (where most optical modes with positive Grüneisen parameters are still not excited) the contribution from the negative Grüneisen parameters will be dominant and thermal expansion coefficient (which is directly proportional to Grüneisen parameters) negative. The lowest negative Grüneisen parameters correspond to the lowest transversal acoustic ZA modes. Phonon frequencies for such modes increase with the in-plane lattice parameter since atoms in the layer upon stretching will be less free to move in the z direction. This is similar to the behaviour of a string, which, when it is stretched, will have vibrations of smaller amplitude and higher frequency. This phenomenon, named "membrane effect", was predicted by Lifshitz in 1952.

Mechanical

The flat graphene sheet is unstable with respect to scrolling i.e. bending into a cylindrical shape, which is its lower-energy state. As of 2009, graphene appeared to be one of the strongest materials known with a breaking strength over 100 times greater than a hypothetical steel film of the same (thin) thickness, with a Young's modulus (stiffness) of 1 TPa (150,000,000 psi). The Nobel announcement illustrated this by saying that a 1 square metre graphene hammock would support a 4 kg cat but would weigh only as much as one of the cat's whiskers, at 0.77 mg (about 0.001% of the weight of 1 m^2 of paper).

However, the process of separating it from graphite, where it occurs naturally, requires technological development to be economical enough to be used in industrial processes. The spring constant of suspended graphene sheets has been measured using an atomic force microscope (AFM). Graphene sheets, held together by van der Waals forces, were suspended over SiO_2 cavities where an AFM tip was probed to test its mechanical properties. Its spring constant was in the range 1–5 N/m and the stiffness was 0.5 TPa, which differs from that of bulk graphite. These high values make graphene very strong and rigid. These intrinsic properties could lead to using graphene for NEMS applications such as pressure sensors and resonators.

As is true of all materials, regions of graphene are subject to thermal and quantum fluctuations in relative displacement. Although the amplitude of these fluctuations is bounded in 3D structures (even in the limit of infinite size), the Mermin–Wagner theorem shows that the amplitude of long-wavelength fluctuations grows logarithmically with the scale of a 2D structure, and would therefore be unbounded in structures of infinite size. Local deformation and elastic strain are negligibly affected by this long-range divergence in relative displacement. It is believed that a sufficiently large 2D structure, in the absence of applied lateral tension, will bend and crumple to form a fluctuating 3D structure. Researchers have observed ripples in suspended layers of graphene, and it has been proposed that the ripples are caused by thermal fluctuations in the material. As a consequence of these dynamical deformations, it is debatable whether graphene is truly a 2D structure.

Spin Transport

Graphene is claimed to be an ideal material for spintronics due to its small spin-orbit interaction and the near absence of nuclear magnetic moments in carbon (as well as a weak hyperfine interaction). Electrical spin current injection and detection has been demonstrated up to room temperature. Spin coherence length above 1 micrometre at room temperature was observed, and control of the spin current polarity with an electrical gate was observed at low temperature.

Strong Magnetic Fields

Graphene's quantum Hall effect in magnetic fields above 10 Teslas or so reveals additional interesting features. Additional plateaus of the Hall conductivity at $\sigma_{xy} = \nu e^2 / h$ with $\nu = 0, \pm1, \pm4$ are observed. Also, the observation of a plateau at $\nu = 3$ and the fractional quantum Hall effect at $\nu = 1/3$ were reported.

These observations with $\nu = 0, \pm1, \pm3, \pm4$ indicate that the four-fold degeneracy (two valley and two spin degrees of freedom) of the Landau energy levels is partially or completely lifted. One hypothesis is that the magnetic catalysis of symmetry breaking is responsible for lifting the degeneracy.

Forms

Nanostripes

Graphene nanoribbons ("nanostripes" in the "zig-zag" orientation), at low temperatures, show spin-polarized metallic edge currents, which also suggests applications in the new field of spintronics. (In the "armchair" orientation, the edges behave like semiconductors.)

Oxide

Using paper-making techniques on dispersed, oxidized and chemically processed graphite in water, the monolayer flakes form a single sheet and create strong bonds. These sheets, called graphene oxide paper have a measured tensile modulus of 32 GPa. The chemical property of graphite oxide is related to the functional groups attached to graphene sheets. These can change the polymerization pathway and similar chemical processes. Graphene oxide flakes in polymers display enhanced photo-conducting properties. Graphene-based membranes are impermeable to all gases and liquids (vacuum-tight). However, water evaporates through them as quickly as if the membrane was not present.

Chemical Modification

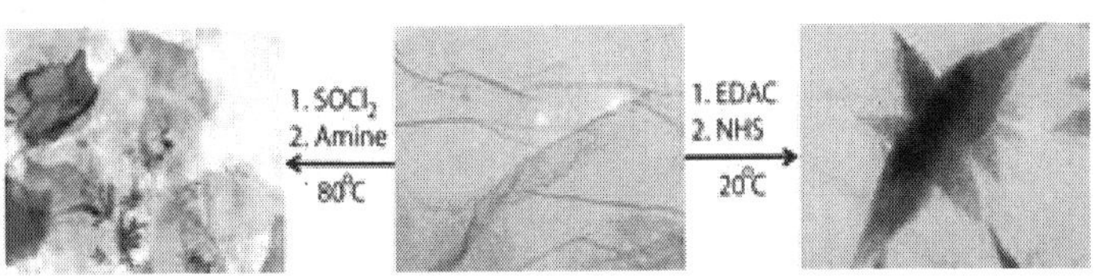

***Figure:** Photograph of single-layer graphene oxide undergoing high temperature chemical treatment, resulting in sheet folding and loss of carboxylic functionality, or through room temperature carbodiimide treatment, collapsing into star-like clusters.*

Soluble fragments of graphene can be prepared in the laboratory through chemical modification of graphite. First, microcrystalline graphite is treated with an acidic mixture of sulfuric acid and nitric acid. A series of oxidation and exfoliation steps produce small graphene plates with carboxyl groups at their edges. These are converted to acid chloride groups by treatment with thionyl chloride; next, they are converted to the corresponding graphene amide via treatment with octadecylamine. The resulting material (circular graphene layers of 5.3 angstrom thickness) is soluble in tetrahydrofuran, tetrachloromethane and dichloroethane.

Refluxing single-layer graphene oxide (SLGO) in solvents leads to size reduction and folding of individual sheets as well as loss of carboxylic group functionality, by up to 20%, indicating thermal instabilities of SLGO sheets dependant on their preparation methodology. When using thionyl chloride, acyl chloride groups result, which can then form aliphatic and aromatic amides with a reactivity conversion of around 70–80%.

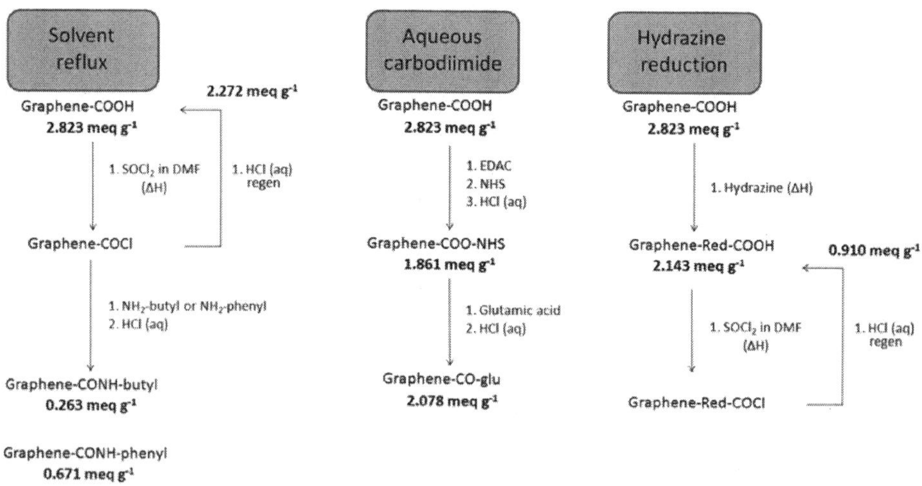

Figure: *Boehm titration results for various chemical reactions of single-layer graphene oxide, which reveal reactivity of the carboxylic groups and the resultant stability of the SLGO sheets after treatment.*

Hydrazine reflux is commonly used for reducing SLGO to SLG(R), but titrations show that only around 20–30% of the carboxylic groups are lost, leaving a significant number available for chemical attachment. Analysis of SLG(R) generated by this route reveals that the system is unstable and using a room temperature stirring with HCl (< 1.0 M) leads to around 60% loss of COOH functionality. Room temperature treatment of SLGO with carbodiimides leads to the collapse of the individual sheets into star-like clusters that exhibited poor subsequent reactivity with amines (ca. 3–5% conversion of the intermediate to the final amide). It is apparent that conventional chemical treatment of carboxylic groups on SLGO generates morphological changes of individual sheets that leads to a reduction in chemical reactivity, which may potentially limit their use in composite synthesis. Therefore, chemical reactions types have been explored. SLGO has also been grafted with polyallylamine, cross-linked through epoxy groups. When filtered into graphene oxide paper, these composites exhibit increased stiffness and strength relative to unmodified graphene oxide paper.

Full hydrogenation from both sides of graphene sheet results in graphane, but partial hydrogenation leads to hydrogenated graphene. Similarly, both-side fluorination of graphene (or chemical and mechanical exfoliation of graphite fluoride) leads to fluorographene (graphene fluoride), while partial fluorination (generally halogenation) provides fluorinated (halogenated) graphene.

Bilayer

Bilayer graphene displays the anomalous quantum Hall effect, a tunable band gap and potential for excitonic condensation –making it a promising candidate for optoelectronic and nanoelectronic applications. Bilayer graphene typically can be found either in twisted configurations where the two layers are rotated relative to each other or graphitic Bernal stacked configurations where half the atoms in one layer lie atop half the atoms in the other. Stacking order and orientation govern the optical and electronic properties of bilayer graphene.

One way to synthesize bilayer graphene is via chemical vapor deposition, which can produce large bilayer regions that almost exclusively conform to a Bernal stack geometry.

Fibre

In 2011, Xinming Li and Hongwei Zhu from Tsinghua University reported a novel yet simple approach to fabricate graphene fibres from chemical vapor deposition grown graphene films. The method was scalable and controllable, delivering tunable morphology and pore structure by controlling the evaporation of solvents with suitable surface tension. Flexible all-solid-state supercapacitors based on this graphene fibres were demonstrated in 2013.

3D

In 2013, a three-dimensional honeycomb of hexagonally arranged carbon was termed 3D graphene, although self-supporting 3D graphene has not yet been produced.

Reinforced

Graphene reinforced with embedded carbon nanotube reinforcing bars ("rebar") is easier to manipulate, while improving the electrical and mechanical qualities of both materials.

Functionalized single- or multiwalled carbon nanotubes are spin-coated on copper foils and then heated and cooled, using the nanotubes themselves as the carbon source. Under heating, the functional carbon groups decompose into graphene, while the nanotubes partially split and form in-plane covalent bonds with the graphene, adding strength. π–π stacking domains add more strength. The nanotubes can overlap, making the material a better conductor than standard CVD-grown graphene. The nanotubes effectively bridge the grain boundaries found in conventional graphene. The technique eliminates the traces of substrate on which later-separated sheets were deposited using epitaxy.

Stacks of a few layers, have been proposed as a cost-effective and physically flexible replacement for indium tin oxide (ITO) used in displays and photovoltaic cells.

Production Techniques

Isolated 2D crystals cannot be grown via chemical synthesis beyond small sizes even in principle, because the rapid growth of phonon density with increasing lateral size forces 2D crystallites to bend into the third dimension. However, other routes to 2d materials exist:

Fundamental forces place seemingly insurmountable barriers in the way of creating [2D crystals]... The nascent 2D crystallites try to minimize their surface energy and inevitably morph into one of the rich variety of stable 3D structures that occur in soot.

But there is a way around the problem. Interactions with 3D structures stabilize 2D crystals during growth. So one can make 2D crystals sandwiched between or placed on top of the atomic planes of a bulk crystal. In that respect, graphene already exists within graphite... One can then hope to fool Nature and extract single-atom-thick crystallites at a low enough temperature that they remain in the quenched state prescribed by the original higher-temperature 3D growth.

The two basic approaches to producing graphene are to cleave multi-layer graphite into single layers or to grow it epitaxially by depositing one layer of carbon onto another material. The former was developed first, using adhesive tape to peel monolayers away. In either case, the graphite must then be bonded to some substrate to retain its 2d shape. Other techniques have also been developed.

Exfoliation

Adhesive Tape: Cleavage is also known as exfoliation. Achieving single layers typically requires multiple exfoliation steps, each producing a slice with fewer layers, until only one remains. Geim and Novosolev used adhesive tape to split their graphene.

After exfoliation the flakes are deposited on a silicon wafer using "dry deposition". Crystallites larger than 1 mm and visible to the naked eye can be obtained with the technique. It is often referred to as a "scotch tape" or "drawing" method. The latter name appeared because the dry deposition resembles drawing with a piece of graphite.

Reduction of Graphite Oxide

Graphite oxide reduction was probably the first method of graphene synthesis. P. Boehm reported producing monolayer flakes of reduced graphene oxide in 1962. Geim acknowledged Boehm's contribution. Rapid heating of graphite oxide and exfoliation yields highly dispersed carbon powder with a few percent of graphene flakes. Reduction of graphite oxide monolayer films, e.g. by hydrazine with annealing in argon/hydrogen, was reported to yield graphene films. However, the quality is lower compared to scotch-tape graphene, due to incomplete removal of functional groups. Furthermore, the oxidation protocol introduces permanent defects due to over-oxidation. The oxidation protocol was enhanced to yield graphene oxide with an almost intact carbon framework that allows efficient removal of functional groups. The measured charge carrier mobility exceeded 1,000 centimetres (393.70 in)/Vs. Spectroscopic analysis of reduced graphene oxide has been conducted.

Sonication

Applying a layer of graphite oxide film to a DVD and burning it in a DVD writer produced a thin graphene film with high electrical conductivity (1738 siemens per metre) and specific surface area (1520 square metres per gram) that was highly resistant and malleable.

Solvent-Aided

Dispersing graphite in a proper liquid medium can produce graphene by sonication. Graphite is separated from graphene by centrifugation, producing graphene concentrations initially up to 0.01 mg/ml in N-methylpyrrolidone (NMP) and later to 2.1 mg/ml in NMP,. Using a suitable ionic liquid as the dispersing liquid medium produced concentrations of 5.33 mg/ml. Graphene concentration produced by this method is very low, because nothing prevents the sheets from restacking due to van der Waals forces. The maximum concentrations achieved are the points at which the van der Waals forces overcome the interactive forces between the graphene sheets and the solvent molecules.

Solvent/Surfactant-Aided

Adding a surfactant to a solvent prior to sonication prevents restacking by adsorbing to the graphene's surface. This produces a higher graphene concentration, but removing the surfactant requires chemical treatments.

Immiscible Liquids

Sonicating graphite at the interface of two immiscible liquids, most notably heptane and water, producing macro-scale graphene films. The graphene sheets are adsorbed to the high energy interface between the heptane and the water, where they are kept from restacking. The graphene remains at the interface even when exposed to force in excess of 300,000 g. The solvents may then be evaporated. The sheets are up to ~95% transparent and conductive.

Epitaxy

Epitaxy refers to the deposition of a crystalline overlayer on a crystalline substrate, where there is registry between the two. In some cases epitaxial graphene layers are coupled to surfaces weakly enough (by Van der Waals forces) to retain the two dimensional electronic band structure of isolated graphene. An example of weakly coupled epitaxial graphene is the one grown on SiC.

Graphene monolayers grown on silicon carbide and iridium are weakly coupled to these substrates (how weakly remains debated) and the graphene–substrate interaction can be further passivated.

Silicon Carbide

Heating silicon carbide (SiC) to high temperatures (>1100 °C) under low pressures (~10^{-6} torr) reduces it to graphene. This process produces epitaxial graphene with dimensions dependent upon the size of the wafer. The face of the SiC used for graphene formation, silicon- or carbon-terminated, highly influences the thickness, mobility and carrier density of the resulting graphene.

Graphene's electronic band-structure (so-called Dirac cone structure) was first visualized in this material. Weak anti-localization is observed in this material, but not in exfoliated graphene produced by the drawing method. Large, temperature-independent mobilities approach those in exfoliated graphene placed on silicon oxide, but lower than mobilities in suspended graphene produced by the drawing method. Even without transfer, graphene on SiC exhibits massless Dirac fermions.

The weak van der Waals force that provides the cohesion of multilayer graphene stacks does not always affect the electronic properties of the individual layers. That is, while the electronic properties of certain multilayered epitaxial graphenes are identical to that of a single layer, in other cases the properties are affected, as they are in bulk graphite. This effect is well understood theoretically and is related to the symmetry of the interlayer interactions.

Epitaxial graphene on SiC can be patterned using standard microelectronics methods. A band gap can be created and tuned by laser irradiation.

Metal Substrates

The atomic structure of a metal substrate can seed the growth of graphene.

Ruthenium

Graphene grown on ruthenium does not typically produce uniform layer thickness. Bonding between the bottom graphene layer and the substrate may affect layer properties.

Iridium

Graphene grown on iridium is very weakly bonded, uniform in thickness and can be highly ordered. As on many other substrates, graphene on iridium is slightly rippled. Due to the long-range order of these ripples, minigaps in the electronic band-structure (Dirac cone) become visible.

Nickel

High-quality sheets of few-layer graphene exceeding 1 cm^2 (0.2 sq in) in area have been synthesized via chemical vapor deposition on thin nickel films with methane as a carbon source. These sheets have been successfully transferred to various substrates.

Graphene has been demonstrated at temperatures compatible with conventional CMOS processing, using a nickel-based alloy with gold as catalyst.

This process dissolves carbon atoms inside a transition metal melt at a certain temperature and then precipitates the dissolved carbon at lower temperatures as single layer graphene (SLG).

The metal is first melted in contact with a carbon source, possibly a graphite crucible inside which the melt is carried out or graphite powder/chunks that are placed in the melt. Keeping the melt in contact with the carbon at a specific temperature dissolves the carbon atoms, saturating the melt based on the binary phase diagram of metal-carbon. Upon lowering the temperature, carbon's solubility decreases and the excess carbon precipitates atop the melt.

The floating layer can be either skimmed or frozen for later removal. Using different morphology, including thick graphite, few layer graphene (FLG) and SLG were observed on metal substrate. Raman spectroscopy proved that SLG had grown on nickel substrate. The SLG Raman spectrum featured no D and D2 band, indicating its pristine nature. Since nickel is not Raman active, direct Raman spectroscopy of graphene layers on top of the nickel is achievable.

Copper

An improvement of this technique employs copper foil; at very low pressure, the growth of graphene automatically stops after a single graphene layer forms. Arbitrarily large films can be created. The single layer growth is also due to the low concentration of carbon in methane. Larger hydrocarbons such as ethane and propane produce bilayer coatings. Atmospheric pressure CVD growth produces multilayer graphene on copper (similar to nickel).

Sodium Ethoxide Pyrolysis

Gram-quantities of graphene were produced by the reduction of ethanol by sodium metal, followed by pyrolysis of the ethoxide product and washing with water to remove sodium salts.

Silicon/Germanium/Hydrogen

A normal silicon wafer coated with a layer of germanium (Ge) dipped in dilute hydrofluoric acid strips the naturally forming germanium oxide groups, creating hydrogen-terminated germanium. Chemical vapor deposition deposits a layer of graphene on top. The graphene can be peeled from the wafer using a dry process and is then ready for use. The wafer can be reused. The graphene is wrinkle-free, high quality and low in defects.

Nanotube Slicing

Graphene can be created by cutting open carbon nanotubes. In one such method multi-walled carbon nanotubes are cut open in solution by action of potassium permanganate and sulfuric acid. In another method graphene nanoribbons were produced by plasma etching of nanotubes partly embedded in a polymer film.

Carbon Dioxide Reduction

A highly exothermic reaction combusts magnesium in an oxidation-reduction reaction with carbon dioxide, producing a variety of carbon nanoparticles including graphene and fullerenes. The carbon dioxide reactant may be either solid (dry-ice) or gaseous. The products of this reaction are carbon and magnesium oxide. US patent 8377408 was issued for this process.

Spin Coating

In 2014, carbon nanotube-reinforced graphene was made via spin coating and annealing functionalized carbon nanotubes. The resulting material was stronger, flexible and more conductive than conventional graphene.

Applications

While as of 2014, graphene is not used in commercial applications, many have been proposed and/or are under active development, in areas including electronics, biological engineering, filtration, lightweight/ strong composite materials, photovoltaics and energy storage.

6

Nanotechnology for Carbon Dioxide Capture

Carbon capture and storage (CCS) (or carbon capture and sequestration), is the process of capturing waste carbon dioxide (CO_2) from large point sources, such as fossil fuel power plants, transporting it to a storage site, and depositing it where it will not enter the atmosphere, normally an underground geological formation.

The aim is to prevent the release of large quantities of CO_2 into the atmosphere (from fossil fuel use in power generation and other industries). It is a potential means of mitigating the contribution of fossil fuel emissions to global warming and ocean acidification.

Although CO_2 has been injected into geological formations for several decades for various purposes, including enhanced oil recovery, the long term storage of CO_2 is a relatively new concept. The first commercial example was Weyburn in 2000. 'CCS' can also be used to describe the scrubbing of CO_2 from ambient air as a geoengineering technique.

An integrated pilot-scale CCS power plant was to begin operating in September 2008 in the eastern German power plant Schwarze Pumpe run by utility Vattenfall, in the hope of answering questions about technological feasibility and economic efficiency. CCS applied to a modern conventional power plant could reduce CO_2 emissions to the atmosphere by approximately 80–90% compared to a plant without CCS. The IPCC estimates that the economic potential of CCS could be between 10% and 55% of the total carbon mitigation effort until year 2100.

Capturing and compressing CO_2 may increase the fuel needs of a coal-fired CCS plant by 25–40%. These and other system costs are estimated to increase the cost of the energy produced by 21–91% for purpose built plants. Applying the technology to existing plants would be more expensive especially if they are far from a sequestration site. Recent industry reports suggest that with successful research, development and deployment (RD&D), sequestered coal-based electricity generation in 2025 may cost less than unsequestered coal-based electricity generation today.

Storage of the CO_2 is envisaged either in deep geological formations, or in the form of mineral carbonates. Deep ocean storage is no longer considered feasible because it greatly increases the problem of ocean acidification.

Geological formations are currently considered the most promising sequestration sites. The National Energy Technology Laboratory (NETL) reported that North America has enough storage capacity for more than 900 years worth of carbon dioxide at current production rates. A general problem is that long term predictions about submarine or underground storage security are very difficult and uncertain, and there is still the risk that CO_2 might leak into the atmosphere.

Capture

Capturing CO_2 is probably most effective at point sources, such as large fossil fuel or biomass energy facilities, industries with major CO_2 emissions, natural gas processing, synthetic fuel plants and fossil fuel-based hydrogen production plants. Extraction (recovery) from air is possible, but not very practical. The CO_2 concentration drops rapidly moving away from the point source. The lower concentration increases the amount of mass flow that must be processed (per tonne of carbon dioxide extracted).

Concentrated CO_2 from the combustion of coal in oxygen is relatively pure, and could be directly processed. Impurities in CO_2 streams could have a significant effect on their phase behaviour and could pose a significant threat of increased corrosion of pipeline and well materials. In instances where CO_2 impurities exist and especially with air capture, a scrubbing process would be needed.

Organisms that produce ethanol by fermentation generate cool, essentially pure CO_2 that can be pumped underground. Fermentation produces slightly less CO_2 than ethanol by weight.

Broadly, three different types of technologies for scrubbing exist: post-combustion, pre-combustion, and oxyfuel combustion:

- In *post combustion capture*, the CO_2 is removed after combustion of the fossil fuel — this is the scheme that would be applied to fossil-fuel burning power plants. Here, carbon dioxide is captured from flue gases at power stations or other large point sources. The technology is well understood and is currently used in other industrial applications, although not at the same scale as might be required in a commercial scale power station.
- The technology for *pre-combustion* is widely applied in fertilizer, chemical, gaseous fuel (H_2, CH_4), and power production. In these cases, the fossil fuel is partially oxidized, for instance in a gasifier. The resulting syngas (CO and H_2) is shifted into CO_2 and H_2. The resulting CO_2 can be captured from a relatively pure exhaust stream. The H_2 can now be used as fuel; the carbon dioxide is removed before combustion takes place. There are several advantages and disadvantages when compared to conventional post combustion carbon dioxide capture. The CO_2 is removed after combustion of fossil fuels, but before the flue gas is expanded to atmospheric pressure. This scheme is applied to new fossil fuel burning power plants, or to existing plants where re-powering is an option. The capture before expansion, i.e. from pressurized gas, is standard in almost all industrial CO_2 capture processes, at the same scale as will be required for utility power plants.
- In *oxy-fuel combustion* the fuel is burned in oxygen instead of air. To limit the resulting flame temperatures to levels common during conventional combustion, cooled flue gas is recirculated and injected into the combustion chamber. The flue gas consists of mainly carbon dioxide and water vapour, the latter of which is condensed through cooling. The result is an almost pure carbon dioxide stream that can be transported to the sequestration site and stored. Power plant processes based on oxyfuel combustion are sometimes referred to as "zero emission" cycles, because the CO_2 stored is not a fraction removed from the flue gas stream (as in the cases of pre- and post-combustion capture) but the flue gas stream itself. A certain fraction of the CO_2 generated during combustion will inevitably end up in the condensed water. To warrant the label "zero emission" the water would thus have to be treated or disposed of appropriately. The technique is promising, but the initial air separation step demands a lot of energy.

An alternate method under development is chemical looping combustion (CLC). Chemical looping uses a metal oxide as a solid oxygen carrier. Metal oxide particles react with a solid, liquid or gaseous fuel in a fluidized bed combustor, producing solid metal particles and a mixture of carbon dioxide and water vapor.

The water vapor is condensed, leaving pure carbon dioxide, which can then be sequestered. The solid metal particles are circulated to another fluidized bed where they react with air, producing heat and regenerating metal oxide particles that are recirculated to the fluidized bed combustor. A variant of chemical looping is calcium looping, which uses the alternating carbonation and then calcination of a calcium oxide based carrier as a means of capturing CO_2.

A few engineering proposals have been made for the more difficult task of capturing CO_2 directly from the air, but work in this area is still in its infancy. Capture costs are estimated to be higher than from point sources, but may be feasible for dealing with emissions from diffuse sources such as automobiles and aircraft. The theoretically required energy for air capture is only slightly more than for capture from point sources. The additional costs come from the devices that use the natural air flow. Global Research Technologies demonstrated a pre-prototype of air capture technology in 2007.

Removing CO_2 from the atmosphere is a form of geoengineering by greenhouse gas remediation. Techniques of this type have received widespread media coverage as they offer the promise of a comprehensive solution to global warming if they can be coupled with effective carbon sequestration technologies.

It is more usual to see such techniques proposed for air capture, than for flue gas treatment. Carbon dioxide capture and storage is more commonly proposed on plants burning coal in oxygen extracted from the air, which means the CO_2 is highly concentrated and no scrubbing process is necessary. According to the Wallula Energy Resource Centre in Washington state, by gasifying coal, it is possible to capture approximately 65% of carbon dioxide embedded in it and sequester it in a solid form.

Transport

After capture, the CO_2 would have to be transported to suitable storage sites. This is done by pipeline, which is generally the cheapest form of transport. In 2008, there were approximately 5,800 km of CO_2 pipelines in the United States, used to transport CO_2 to oil production fields where it is then injected into older fields to extract oil. The injection of CO_2 to produce oil is generally called *Enhanced Oil Recovery* or EOR. In addition, there are several pilot programs in various stages to test the long-term storage of CO_2 in non-oil producing geologic formations.

According to the Congressional Research Service, "There are important unanswered questions about pipeline network requirements, economic regulation, utility cost recovery, regulatory classification of CO_2 itself, and pipeline safety. Furthermore, because CO_2 pipelines for enhanced oil recovery are already in use today, policy decisions affecting CO_2 pipelines take on an urgency that is unrecognised by many. Federal classification of CO_2 as both a commodity (by the Bureau of Land Management) and as a pollutant (by the Environmental Protection Agency) could potentially create an immediate conflict which may need to be addressed not only for the sake of future CCS implementation, but also to ensure consistency of future CCS with CO_2 pipeline operations today."

Ships could also be utilised for transport where pipelines are not feasible. These methods are currently used for transporting CO_2 for other applications.

Carbon Dioxide Capture with Nanometric Thin-Film Membranes

The burning of fossil fuels like coal and oil pumps billions of tons of greenhouse gases into the atmosphere annually, changing the world's climate in the process. It is estimated that approximately one third of total world carbon dioxide emissions come from energy production in power plants. One possible option for reducing CO_2 emissions from power plants is to capture them before they hit the atmosphere and store the gas underground. This technique is called Carbon Dioxide Capture and Storage.

However, before CO_2 can be stored, it must be separated from the other waste gases resulting from combustion or industrial processes. Most current methods used for this type of filtration are expensive and require the use of chemicals. Nanotechnology techniques to fabricate nanoscale thin membranes could lead to new membrane technology that could change that.

Current membranes are in many cases not competitive for large scale applications, because their permeance (flux) for carbon dioxide is not high enough; state-of-the-art membranes exhibit a carbon dioxide permeance of about 0.3 m3 (STP) m-2 h-1 bar-1.

Researchers in Germany have now reported the development and manufacturing of nanometric thin film membranes with record performance. By combining a CO_2-philic polymer design with controlled nanofabrication conditions they have achieved defect-free selective layers far thinner than 100 nm. These membranes exhibit unmatched high CO_2 permeance combined with high selectivity and stability.

Researchers from the Institute of Materials Research and Institute of Polymer Research, both at GKSS-Research Centre Geesthacht, have fabricated multiblock copolymers consisting of polyethylene oxide and polybutyleneterephtalate into thin-film composite membranes on laboratory and technical scales.

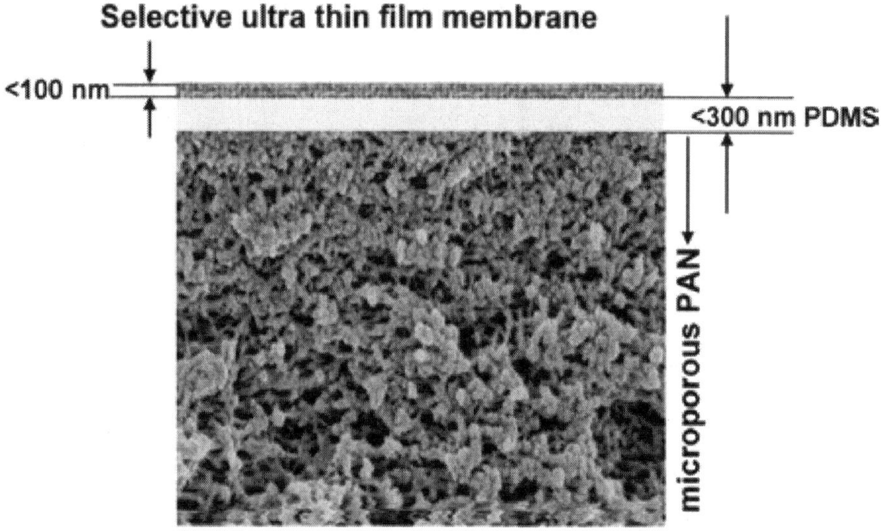

Figure: *Schematic representation of the thin-film membrane.*

In order to manufacture their ultra-thin defect-free coatings with extremely high CO_2 permeance and high selectivity, the researchers used a highly permeable intermediate layer (PDMS; thickness <300 nm) between the microporous support and the selective coating. The highly permeable intermediate layer worked as a gutter and protective coating which prevents the diluted polymer solution penetration into the porous structure, and at the same time renders the entire membrane surface smoother.

To test the performance of their membranes, the GKSS researchers performed experiments with gas mixtures in order to evaluate the application in power plants. They found that their ultra-thin film membrane has roughly a 20 times higher CO_2 permeance than commercial membranes, combined with one of the highest CO_2/N2-selectivities known (facilitated transport materials excluded).

According to the team, a preliminary technical and economical analysis shows that this membrane has potential for carbon capture in coal-fired power plants. They note that the capture cost using the conventional amine absorption process is in the range of 30-50 Euro per tonne recovered CO_2.

By applying a membrane-based process design the separation costs can be lower than 30 Euro/t using the high permeance membrane developed at GKSS.

These membranes are currently under evaluation at pilot scale in a German project for CO_2 capture in power plants.

Nanotechnology in Dye Sensitized

A dye-sensitized solar cell (DSSC, DSC or DYSC) is a low-cost solar cell belonging to the group of thin film solar cells. It is based on a semiconductor formed between a photo-sensitized anode and an electrolyte, a *photoelectrochemical* system. The modern version of a dye solar cell, also known as the Grätzel cell, was originally co-invented in 1988 by Brian O'Regan and Michael Grätzel at UC Berkeley and this work was later developed by the aforementioned scientists at the École Polytechnique Fédérale de Lausanne until the publication of the first high efficiency DSSC in 1991. Michael Grätzel has been awarded the 2010 Millennium Technology Prize for this invention.

***Figure:** A selection of dye-sensitized solar cells*

The DSSC has a number of attractive features; it is simple to make using conventional roll-printing techniques, is semi-flexible and semi-transparent which offers a variety of uses not applicable to glass-based systems, and most of the materials used are low-cost.

In practice it has proven difficult to eliminate a number of expensive materials, notably platinum and ruthenium, and the liquid electrolyte presents a serious challenge to making a cell suitable for use in all weather. Although its conversion efficiency is less than the best thin-film cells, in theory its price/performance ratio should be good enough to allow them to compete with fossil fuel electrical generation by achieving grid parity. Commercial applications, which were held up due to chemical stability problems, are now forecast in the European Union Photovoltaic Roadmap to significantly contribute to renewable electricity generation by 2020.

Current Technology: Semiconductor Solar Cells

In a traditional solid-state semiconductor, a solar cell is made from two doped crystals, one doped with n-type impurities (n-type semiconductor), which add additional free conduction band electrons, and the other doped with p-type impurities (p-type semiconductor), which add additional electron holes. When placed in contact, some of the electrons in the n-type portion flow into the p-type to "fill in" the missing electrons, also known as electron holes. Eventually enough electrons will flow across the boundary to equalize the Fermi levels of the two materials. The result is a region at the interface, the p-n junction, where charge carriers are depleted and/or accumulated on each side of the interface. In silicon, this transfer of electrons produces a potential barrier of about 0.6 to 0.7 V.

When placed in the sun, photons of the sunlight can excite electrons on the p-type side of the semiconductor, a process known as photoexcitation. In silicon, sunlight can provide enough energy to push an electron out of the lower-energy valence band into the higher-energy conduction band. As the name implies, electrons in the conduction band are free to move about the silicon. When a load is placed across the cell as a whole, these electrons will flow out of the p-type side into the n-type side, lose energy while moving through the external circuit, and then flow back into the p-type material where they can once again re-combine with the valence-band hole they left behind. In this way, sunlight creates an electrical current.

In any semiconductor, the band gap means that only photons with that amount of energy, or more, will contribute to producing a current.

In the case of silicon, the majority of visible light from red to violet has sufficient energy to make this happen. Unfortunately higher energy photons, those at the blue and violet end of the spectrum, have more than enough energy to cross the band gap; although some of this extra energy is transferred into the electrons, the majority of it is wasted as heat.

Another issue is that in order to have a reasonable chance of capturing a photon, the n-type layer has to be fairly thick. This also increases the chance that a freshly ejected electron will meet up with a previously created hole in the material before reaching the p-n junction. These effects produce an upper limit on the efficiency of silicon solar cells, currently around 12 to 15% for common modules and up to 25% for the best laboratory cells (About 30% is the theoretical maximum efficiency for single band gap solar cells).

By far the biggest problem with the conventional approach is cost; solar cells require a relatively thick layer of doped silicon in order to have reasonable photon capture rates, and silicon processing is expensive. There have been a number of different approaches to reduce this cost over the last decade, notably the thin-film approaches, but to date they have seen limited application due to a variety of practical problems. Another line of research has been to dramatically improve efficiency through the multi-junction approach, although these cells are very high cost and suitable only for large commercial deployments. In general terms the types of cells suitable for rooftop deployment have not changed significantly in efficiency, although costs have dropped somewhat due to increased supply.

Dye-Sensitized Solar Cells

In the late 1960s it was discovered that illuminated organic dyes can generate electricity at oxide electrodes in electrochemical cells. In an effort to understand and simulate the primary processes in photosynthesis the phenomenon was studied at the University of California at Berkeley with chlorophyll extracted from spinach (bio-mimetic or bionic approach). On the basis of such experiments electric power generation via the dye sensitization solar cell (DSSC) principle was demonstrated and discussed in 1972. The instability of the dye solar cell was identified as a main challenge. Its efficiency could, during the following two decades, be improved by optimizing the porosity of the electrode prepared from fine oxide powder, but the instability remained a problem. A modern DSSC is composed of a porous layer of titanium dioxide nanoparticles, covered with a molecular dye that absorbs sunlight, like the chlorophyll in green leaves. The titanium dioxide is immersed under an electrolyte solution, above which is a platinum-based catalyst. As in a conventional alkaline battery, an anode (the titanium dioxide) and a cathode (the platinum) are placed on either side of a liquid conductor (the electrolyte).

Sunlight passes through the transparent electrode into the dye layer where it can excite electrons that then flow into the titanium dioxide. The electrons flow toward the transparent electrode where they are collected for powering a load. After flowing through the external circuit, they are re-introduced into the cell on a metal electrode on the back, flowing into the electrolyte. The electrolyte then transports the electrons back to the dye molecules.

Dye-sensitized solar cells separate the two functions provided by silicon in a traditional cell design. Normally the silicon acts as both the source of photoelectrons, as well as providing the electric field to separate the charges and create a current. In the dye-sensitized solar cell, the bulk of the semiconductor is used solely for charge transport, the photoelectrons are provided from a separate photosensitive dye. Charge separation occurs at the surfaces between the dye, semiconductor and electrolyte.

The dye molecules are quite small (nanometre sized), so in order to capture a reasonable amount of the incoming light the layer of dye molecules needs to be made fairly thick, much thicker than the molecules themselves. To address this problem, a nanomaterial is used as a scaffold to hold large numbers of the dye molecules in a 3-D matrix, increasing the number of molecules for any given surface area of cell. In existing designs, this scaffolding is provided by the semiconductor material, which serves double-duty.

Construction

In the case of the original Grätzel and O'Regan design, the cell has 3 primary parts. On top is a transparent anode made of fluoride-doped tin dioxide (SnO_2:F) deposited on the back of a (typically glass) plate. On the back of this conductive plate is a thin layer of titanium dioxide (TiO_2), which forms into a highly porous structure with an extremely high surface area. TiO_2 only absorbs a small fraction of the solar photons (those in the UV). The plate is then immersed in a mixture of a photosensitive ruthenium-polypyridine dye (also called molecular sensitizers) and a solvent. After soaking the film in the dye solution, a thin layer of the dye is left covalently bonded to the surface of the TiO_2.

A separate plate is then made with a thin layer of the iodide electrolyte spread over a conductive sheet, typically platinum metal. The two plates are then joined and sealed together to prevent the electrolyte from leaking. The construction is simple enough that there are hobby kits available to hand-construct them. Although they use a number of "advanced" materials, these are inexpensive compared to the silicon needed for normal cells because they require no expensive manufacturing steps. TiO_2, for instance, is already widely used as a paint base.

One of the efficient DSSCs devices uses ruthenium-based molecular dye, e.g. [Ru(4,40-dicarboxy-2,20-bipyridine)2(NCS)2] (N3), that is bound to a photoanode via carboxylate moieties. The photoanode consists of 12 μm thick ûlm of transparent 10–20 nm diameter TiO_2 nanoparticles covered with a 4 μm thick ûlm of much larger (400 nm diameter) particles that scatter photons back into the transparent ûlm. The excited dye rapidly injects an electron into the TiO_2 after light absorption. The injected electron diffuses through the sintered particle network to be collected at the front side transparent conducting oxide (TCO) electrode, while the dye is regenerated via reduction by a redox shuttle, I_3/I, dissolved in a solution. Diffusion of the oxidized form of the shuttle to the counter electrode completes the circuit.

Mechanism of DSSCs

The main processes that occur in a DSSC

Step 1:The following primary steps convert photons to current:

1. The incident photon is absorbed by Ru complex photosensitizers adsorbed on the TiO_2 surface.
2. The photosensitizers are excited from the ground state (S) to the excited state (S^*). The excited electrons are injected into the conduction band of the TiO_2 electrode. This results in the oxidation of the photosensitizer (S^+).

   $S + h\nu \rightarrow S^*$ (1)

   $S^* \rightarrow S^+ + e^-$ (TiO_2) (2)

3. The injected electrons in the conduction band of TiO_2 are transported between TiO_2 nanoparticles with diffusion toward the back contact (TCO). And the electrons finally reach the counter electrode through the circuit.
4. The oxidized photosensitizer (S^+) accepts electrons from the I^- ion redox mediator leading to regeneration of the ground state (S), and the I^- is oxidized to the oxidized state, I_3^-.

   $S^+ + e^- \rightarrow S$ (3)

5. The oxidized redox mediator, I_3^-, diffuses toward the counter electrode and then it is reduced to I^- ions.

$$I_3^- + 2\ e^- \rightarrow 3\ I^-\ (4)$$

The efficiency of a DSSC is depends on four energy levels of the component: the excited state (approximately LUMO) and the ground state (HOMO) of the photosensitizer, the Fermi level of the TiO_2 electrode and the redox potential of the mediator (I^-/I_3^-) in the electrolyte.

Nanoplant-Like Morphology

In DSSC, electrodes consisted of sintered semiconducting nanoparticles, mainly TiO_2 or ZnO. These nanoparticle DSSCs rely on trap-limited diffusion through the semiconductor nanoparticles for the electron transport.

This limits the device efficiency since it is a slow transport mechanism. Recombination is more likely to occur at longer wavelengths of radiation. Moreover sintering of nanoparticles requires a high temperature of about 450 °C, which restricts the fabrication of these cells to robust, rigid solid substrates. It has been proved that there is in increase in the efficiency of DSSC, if the sintered nanoparticle electrode is replaced by a specially designed electrode possessing an exotic 'nanoplant-like' morphology.

Operation

Sunlight enters the cell through the transparent SnO_2:F top contact, striking the dye on the surface of the TiO_2. Photons striking the dye with enough energy to be absorbed create an excited state of the dye, from which an electron can be "injected" directly into the conduction band of the TiO_2. From there it moves by diffusion (as a result of an electron concentration gradient) to the clear anode on top.

Meanwhile, the dye molecule has lost an electron and the molecule will decompose if another electron is not provided.

The dye strips one from iodide in electrolyte below the TiO_2, oxidizing it into triiodide. This reaction occurs quite quickly compared to the time that it takes for the injected electron to recombine with the oxidized dye molecule, preventing this recombination reaction that would effectively short-circuit the solar cell.

The triiodide then recovers its missing electron by mechanically diffusing to the bottom of the cell, where the counter electrode re-introduces the electrons after flowing through the external circuit.

Efficiency

Several important measures are used to characterize solar cells. The most obvious is the total amount of electrical power produced for a given amount of solar power shining on the cell. Expressed as a percentage, this is known as the *solar conversion efficiency*. Electrical power is the product of current and voltage, so the maximum values for these measurements are important as well, J_{sc} and V_{oc} respectively. Finally, in order to understand the underlying physics, the "quantum efficiency" is used to compare the chance that one photon (of a particular energy) will create one electron.

In quantum efficiency terms, DSSCs are extremely efficient. Due to their "depth" in the nanostructure there is a very high chance that a photon will be absorbed, and the dyes are very effective at converting them to electrons. Most of the small losses that do exist in DSSC's are due to conduction losses in the TiO_2 and the clear electrode, or optical losses in the front electrode. The overall quantum efficiency for green light is about 90%, with the "lost" 10% being largely accounted for by the optical losses in top electrode. The quantum efficiency of traditional designs vary, depending on their thickness, but are about the same as the DSSC.

In theory, the maximum voltage generated by such a cell is simply the difference between the (*quasi*-)Fermi level of the TiO_2 and the redox potential of the electrolyte, about 0.7 V under solar illumination conditions (V_{oc}). That is, if an illuminated DSSC is connected to a voltmetre in an "open circuit", it would read about 0.7 V. In terms of voltage, DSSCs offer slightly higher V_{oc} than silicon, about 0.7 V compared to 0.6 V. This is a fairly small difference, so real-world differences are dominated by current production, J_{sc}.

Although the dye is highly efficient at converting absorbed photons into free electrons in the TiO_2, only photons absorbed by the dye ultimately produce current. The rate of photon absorption depends upon the absorption spectrum of the sensitized TiO_2 layer and upon the solar flux spectrum. The overlap between these two spectra determines the maximum possible photocurrent. Typically used dye molecules generally have poorer absorption in the red part of the spectrum compared to silicon, which means that fewer of the photons in sunlight are usable for current generation. These factors limit the current generated by a DSSC, for comparison, a traditional silicon-based solar cell offers about 35 mA/cm^2, whereas current DSSCs offer about 20 mA/cm^2.

Overall peak power conversion efficiency for current DSSCs is about 11%. Current record for prototypes lies at 15%.

Degradation

DSSCs degrade when exposed to ultraviolet radiation. The barrier layer may include UV stabilizers and/or UV absorbing luminescent chromophores (which emit at longer wavelengths) and antioxidants to protect and improve the efficiency of the cell.

Advantages

DSSCs are currently the most efficient third-generation (2005 Basic Research Solar Energy Utilisation 16) solar technology available. Other thin-film technologies are typically between 5% and 13%, and traditional low-cost commercial silicon panels operate between 14% and 17%. This makes DSSCs attractive as a replacement for existing technologies in "low density" applications like rooftop solar collectors, where the mechanical robustness and light weight of the glass-less collector is a major advantage. They may not be as attractive for large-scale deployments where higher-cost higher-efficiency cells are more viable, but even small increases in the DSSC conversion efficiency might make them suitable for some of these roles as well.

There is another area where DSSCs are particularly attractive. The process of injecting an electron directly into the TiO_2 is qualitatively different to that occurring in a traditional cell, where the electron is "promoted" within the original crystal. In theory, given low rates of production, the high-energy electron in the silicon could re-combine with its own hole, giving off a photon (or other form of energy) and resulting in no current being generated. Although this particular case may not be common, it is fairly easy for an electron generated in another molecule to hit a hole left behind in a previous photoexcitation.

In comparison, the injection process used in the DSSC does not introduce a hole in the TiO_2, only an extra electron. Although it is energetically possible for the electron to recombine back into the dye, the rate at which this occurs is quite slow compared to the rate that the dye regains an electron from the surrounding electrolyte. Recombination directly from the TiO_2 to species in the electrolyte is also possible although, again, for optimized devices this reaction is rather slow. On the contrary, electron transfer from the platinum coated electrode to species in the electrolyte is necessarily very fast.

As a result of these favourable "differential kinetics", DSSCs work even in low-light conditions. DSSCs are therefore able to work under cloudy skies and non-direct sunlight, whereas traditional designs would suffer a "cutout" at some lower limit of illumination, when charge carrier mobility is low and recombination becomes a major issue. The cutoff is so low they are even being proposed for indoor use, collecting energy for small devices from the lights in the house.

A practical advantage, one DSSCs share with most thin-film technologies, is that the cell's mechanical robustness indirectly leads to higher efficiencies in higher temperatures. In any semiconductor, increasing temperature will promote some electrons into the conduction band "mechanically". The fragility of traditional silicon cells requires them to be protected from the elements, typically by encasing them in a glass box similar to a greenhouse, with a metal backing for strength. Such systems suffer noticeable decreases in efficiency as the cells heat up internally. DSSCs are normally built with only a thin layer of conductive plastic on the front layer, allowing them to radiate away heat much easier, and therefore operate at lower internal temperatures.

Disadvantages

The major disadvantage to the DSSC design is the use of the liquid electrolyte, which has temperature stability problems. At low temperatures the electrolyte can freeze, ending power production and potentially leading to physical damage. Higher temperatures cause the liquid to expand, making sealing the panels a serious problem. Another disadvantage is that costly ruthenium (dye), platinum (catalyst) and conducting glass or plastic (contact) are needed to produce a DSSC. A third major drawback is that the electrolyte solution contains volatile organic compounds (or VOC's), solvents which must be carefully sealed as they are hazardous to human health and the environment. This, along with the fact that the solvents permeate plastics, has precluded large-scale outdoor application and integration into flexible structure.

Replacing the liquid electrolyte with a solid has been a major ongoing field of research. Recent experiments using solidified melted salts have shown some promise, but currently suffer from higher degradation during continued operation, and are not flexible.

Photocathodes and Tandem Cells

Dye sensitised solar cells operate as a photoanode (n-DSC), where photocurrent result from electron injection by the sensitized dye. Photocathodes (p-DSCs) operate in an inverse mode compared to the conventional n-DSC, where dye-excitation is followed by rapid electron transfer from a p-type semiconductor to the dye (dye-sensitized hole injection, instead of electron injection). Such p-DSCs and n-DSCs can be combined to construct tandem solar cells (pn-DSCs) and the theoretical efficiency of tandem DSCs is well beyond that of single-junction DSCs.

A standard tandem cell consists of one n-DSC and one p-DSC in a simple sandwich configuration with an intermediate electrolyte layer. n-DSC and p-DSC are connected in series, which implies that the resulting photocurrent will be controlled by the weakest photoelectrode, whereas photovoltages are additive. Thus, photocurrent matching is very important for the construction of highly efficient tandem pn-DSCs. However, unlike n-DSCs, fast charge recombination following dye-sensitized hole injection usually resulted in low photocurrents in p-DSC and thus hampered the efficiency of the overall device.

Researchers have found that using dyes comprising a perylenemonoimid (PMI) as the acceptor and an oligothiophene coupled to triphenylamine as the donor greatly improve the performance of p-DSC by reducing charge recombination rate following dye-sensitized hole injection. The researchers constructed a tandem DSC device with NiO on the p-DSC side and TiO_2 on the n-DSC side. Photocurrent matching was achieved through adjustment of NiO and TiO_2 film thicknesses to control the optical absorptions and therefore match the photocurrents of both electrodes. The energy conversion efficiency of the device is 1.91%, which exceeds the efficiency of its individual components, but is still much lower than that of high performance n-DSC devices (6%–11%). The results are still promising since the tandem DSC was in itself rudimentary. The dramatic improvement in performance in p-DSC can eventually lead to tandem devices with much greater efficiency than lone n-DSCs.

Development

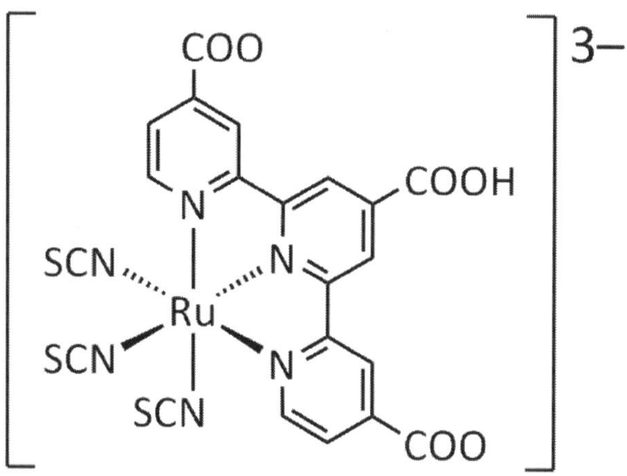

Figure: *"Black Dye", an anionic Ru-terpyridine complex*

The dyes used in early experimental cells (circa 1995) were sensitive only in the high-frequency end of the solar spectrum, in the UV and blue. Newer versions were quickly introduced (circa 1999) that had much wider frequency response, notably "triscarboxy-ruthenium terpyridine" [Ru(4,4',4"-$(COOH)_3$-terpy)$(NCS)_3$], which is efficient right into the low-frequency range of red and IR light. The wide spectral response results in the dye having a deep brown-black colour, and is referred to simply as "black dye". The dyes have an excellent chance of converting a photon into an electron, originally around 80% but improving to almost perfect conversion in more recent dyes, the overall efficiency is about 90%, with the "lost" 10% being largely accounted for by the optical losses in top electrode.

A solar cell must be capable of producing electricity for at least twenty years, without a significant decrease in efficiency (life span). The "black dye" system was subjected to 50 million cycles, the equivalent of ten years' exposure to the sun in Switzerland. No discernible performance decrease was observed. However the dye is subject to breakdown in high-light situations. Over the last decade an extensive research program has been carried out to address these concerns. The newer dyes included 1-ethyl-3 methylimidazolium tetrocyanoborate [EMIB$(CN)_4$] which is extremely light- and temperature-stable, copper-diselenium [Cu(In,GA)Se_2] which offers higher conversion efficiencies, and others with varying special-purpose properties.

DSSCs are still at the start of their development cycle. Efficiency gains are possible and have recently started more widespread study. These include the use of quantum dots for conversion of higher-energy (higher frequency) light into multiple electrons, using solid-state electrolytes for better temperature response, and changing the doping of the TiO_2 to better match it with the electrolyte being used.

New Developments

2003: A group of researchers at the Swiss Federal Institute of Technology has reportedly increased the thermostability of DSC by using amphiphilic ruthenium sensitizer in conjunction with quasi-solid-state gel electrolyte. The stability of the device matches that of a conventional inorganic silicon based solar cell. The cell sustained heating for 1,000 h at 80 °C.

The group has previously prepared a ruthenium amphiphilic dye Z-907 (cis-Ru(H_2dcbpy)(dnbpy)$(NCS)_2$,where the ligand H_2dcbpy is 4,42 -dicarboxylic acid-2,22 -bipyridine and dnbpy is 4,42 -dinonyl-2,22 -

bipyridine) to increase dye tolerance to water in the electrolytes. In addition, the group also prepared a quasi-solid-state gel electrolye with a 3-methoxypropionitrile (MPN)-based liquid electrolyte that was solidified by a photochemically stable fluorine polymer, poly(vinylidenefluoride-co-hexafluoropropylene (PVDF-HFP).

The use of the amphiphilic Z-907 dye in conjunction with the polymer gel electrolyte in DSC achieved an energy conversion efficiency of 6.1%. More importantly, the device was stable under thermal stress and soaking with light. The high conversion efficiency of the cell was sustained after heating for 1,000 h at 80 °C, maintaining 94% of its initial value. After accelerated testing in a solar simulator for 1,000 h of light-soaking at 55 °C (100 mW cm^{-2}) the efficiency had decreased by less than 5% for cells covered with an ultraviolet absorbing polymer film. These results are well within the limit for that of traditional inorganic silicon solar cells.

The enhanced performance may arise from a decrease in solvent permeation across the sealant due to the application of the polymer gel electrolyte. The polymer gel electrolyte is quasi-solid at room temperature, and becomes a viscous liquid (viscosity: 4.34 mPa• s) at 80 °C compared with the traditional liquid electrolyte (viscosity: 0.91 mPa ·s). The much improved stabilities of the device under both thermal stress and soaking with light has never before been seen in DSCs, and they match the durability criteria applied to solar cells for outdoor use, which makes these devices viable for practical application.

2006: The first successful solid-hybrid dye-sensitized solar cells were reported.

To improve electron transport in these solar cells, while maintaining the high surface area needed for dye adsorption, two researchers have designed alternate semiconductor morphologies, such as arrays of nanowires and a combination of nanowires and nanoparticles, to provide a direct path to the electrode via the semiconductor conduction band. Such structures may provide a means to improve the quantum efficiency of DSSCs in the red region of the spectrum, where their performance is currently limited.

On August 2006, to prove the chemical and thermal robustness of the 1-ethyl-3 methylimidazolium tetracyanoborate solar cell, the researchers subjected the devices to heating at 80 °C in the dark for 1000 hours, followed by light soaking at 60 °C for 1000 hours. After dark heating and light soaking, 90% of the initial photovoltaic efficiency was maintained – the first time such excellent thermal stability has been observed for a liquid electrolyte that exhibits such a high conversion efficiency. Contrary to silicon solar cells, whose performance declines with increasing temperature, the dye-sensitized solar-cell devices were only negligibly influenced when increasing the operating temperature from ambient to 60 °C.

April 2007: Wayne Campbell at Massey University, New Zealand, has experimented with a wide variety of organic dyes based on porphyrin. In nature, porphyrin is the basic building block of the hemoproteins, which include chlorophyll in plants and hemoglobin in animals. He reports efficiency on the order of 5.6% using these low-cost dyes.

June 2008: An article published in *Nature Materials* demonstrated cell efficiencies of 8.2% using a new solvent-free liquid redox electrolyte consisting of a melt of three salts, as an alternative to using organic solvents as an electrolyte solution. Although the efficiency with this electrolyte is less than the 11% being delivered using the existing iodine-based solutions, the team is confident the efficiency can be improved.

2009: A group of researchers at Georgia Tech made dye-sensitized solar cells with a higher effective surface area by wrapping the cells around a quartz optical fibre. The researchers removed the cladding from optical fibres, grew zinc oxide nanowires along the surface, treated them with dye molecules, surrounded the fibres by an electrolyte and a metal film that carries electrons off the fibre. The cells are six times more efficient than a zinc oxide cell with the same surface area. Photons bounce inside the fibre as they travel, so there are more chances to interact with the solar cell and produce more current. These devices only collect light at the tips, but future fibre cells could be made to absorb light along the entire length of the fibre, which would require a coating that is conductive as well as transparent. Max Shtein of the University of Michigan said a sun-tracking system would not be necessary for such cells, and would work on cloudy days when light is diffuse.

2010: Researchers at the École Polytechnique Fédérale de Lausanne and at the Université du Québec à Montréal claim to have overcome two of the DSC's major issues:

- "New molecules" have been created for the electrolyte, resulting in a liquid or gel that is transparent and non-corrosive, which can increase the photovoltage and improve the cell's output and stability.

- At the cathode, platinum was replaced by cobalt sulfide, which is far less expensive, more efficient, more stable and easier to produce in the laboratory.

2011: Dyesol and Tata Steel Europe announced in June the development of the world's largest dye sensitized photovoltaic module, printed onto steel in a continuous line.

Dyesol and CSIRO announced in October a Successful Completion of Second Milestone in Joint Dyesol / CSIRO Project. Dyesol Director Gordon Thompson said, "The materials developed during this joint collaboration have the potential to significantly advance the commercialisation of DSC in a range of applications where performance and stability are essential requirements. Dyesol is extremely encouraged by the breakthroughs in the chemistry allowing the production of the target molecules. This creates a path to the immediate commercial utilisation of these new materials."

Dyesol and Tata Steel Europe announced in November the targeted development of Grid Parity Competitive BIPV solar steel that does not require government subsidised feed in tariffs. TATA-Dyesol "Solar Steel" Roofing is currently being installed on the Sustainable Building Envelope Centre (SBEC) in Shotton, Wales.

2012: Northwestern University researchers announced a solution to a primary problem of DSSCs, that of difficulties in using and containing the liquid electrolyte and the consequent relatively short useful life of the device. This is achieved through the use of nanotechnology and the conversion of the liquid electrolyte to a solid. The current efficiency is about half that of silicon cells, but the cells are lightweight and potentially of much lower cost to produce.

2013: During the last 5-10 years, a new kind of DSSC has been developed - the solid state dye-sensitized solar cell. In this case the liquid electrolyte is replaced by one of several solid hole conducting materials. From 2009 to 2013 the efficiency of Solid State DSSCs has dramatically increased from 4% to 15%. Michael Graetzel announced the fabrication of Solid State DSSCs with 15.0% efficiency, reached by the means of a hybrid perovskite $CH_3NH_3PbI_3$ dye, subsequently deposited from the separated solutions of CH_3NH_3I and PbI_2.

First architectural integration at EPFL's new convention centre, in partnership with Romande Energie. The total surface will be 300 square metres, in 1400 modules of 50 cm x 35 cm. Designed by artists Daniel Schlaepfer and Catherine Bolle.

Market

Several commercial providers are promising availability of DSCs in the near future:

- Dyesol officially opened its new manufacturing facilities in Queanbeyan Australia on 7 October 2008. It has subsequently announced partnerships with Tata Steel (TATA-Dyesol) and Pilkington Glass (Dyetec-Solar) for the development and large scale manufacture of DSC BIPV. Dyesol has also entered working relationships with Merck, Umicore, CSIRO, Japanese Ministry of Economy and Trade, Singapore Aerospace Manufacturing and a joint Venture with TIMO Korea (Dyesol-TIMO).
- Solaronix, a Swiss company specialised in the production of DSC materials since 1993, has extended their premises in 2010 to host a manufacturing pilot line of DSC modules.
- SolarPrint founded in 2008 by Dr. Mazhar Bari, Andre Fernon and Roy Horgan. SolarPrint is the first Ireland-based commercial entity involved in the manufacturing of PV technology. SolarPrint's innovation is the solution to the solvent based electrolyte which to date has prohibited the mass commercialisation of DSSC.
- G24innovations founded in 2006, based in Cardiff, South Wales, UK. On 17 October 2007, claimed the production of the first commercial grade dye sensitised thin films.
- Sony Corporation has developed dye-sensitized solar cells with an energy conversion efficiency of 10%, a level seen as necessary for commercial use.
- Tasnee Enters Strategic Investment Agreement with Dyesol.

Nanocatalysis for Fuel Production

Nanomaterial-based catalysts are usually heterogeneous catalysts broken up into metal nanoparticles in order to speed up the catalytic process. Metal nanoparticles have a higher surface area so there is increased catalytic activity because more catalytic reactions can occur at the same time. Nanoparticle catalysts can also be easily separated and recycled with more retention of catalytic activity than their bulk counterparts. These catalysts can play two different roles in catalytic processes: they can be the site of catalysis or they can act as a support for catalytic processes. They are typically used under mild conditions to prevent decomposition of the nanoparticles at extreme conditions.

Functionalized metal nanoparticles are more stable in solution compared to non-functionalized metal nanoparticles. In liquid solutions, the metal nanoparticles are close enough together to be affected by van der Waals force. If there isn't anything to oppose these forces, then the nanoparticles will aggregate, which will lead to a decrease in catalytic activity by lowering the surface area.

For organometallic functionalized nanoparticles, ligands are coordinated to the metal centre to prevent aggregation. Using different ligands alters the properties and sizes of the nanoparticle catalysts. Nanoparticles can also be functionalized with polymers or oligomers to sterically stabilize the nanoparticles by providing a protective layer that prevents the nanoparticles from interacting with each other. Alloys of two metals, called bimetallic nanoparticles, are used to create synergistic effects on catalysis between the two metals.

Applications

Dehalogenation and Hydrogenation: Nanoparticle catalysts can be used in the hydrogenolysis of C-Cl bonds such as polychlorinated biphenyls. Hydrogenation of halogenated aromatic amines is also important for the synthesis of herbicides and pesticides as well as diesel fuel. In organic chemistry, hydrogenation of a C-Cl bond with deuterium is used to selectively label the aromatic ring for use in experiments dealing with the kinetic isotope effect. Buil *et al.* created rhodium complexes that generated rhodium nanoparticles. These nanoparticles catalyzed the dehalogenation of aromatic compounds as well as the hydrogenation of benzene to cyclohexane. Polymer-stabilized nanoparticles can also be used for the hydrogenation of cinnamaldehyde and citronellal. Yu *et al.* found that the ruthenium nanocatalysts are more selective in the hydrogenation of citronellal compared to the traditional catalysts used.

Hydrosilylation Reactions

***Figure:** Hydrosilylation reaction*

The Reduction of gold, cobalt, nickel, palladium, or platinum organometallic complexes with silanes creates a catalytically active metal nanoparticle that catalyzes the hydrosilylation reaction, which is important for the synthesis of optically active alcohols. BINAP functionalized palladium nanoparticles and gold nanoparticles have been used for the hydrosilylaytion of styrene under mild conditions;

they were found to be more catalytically active and more stable than non-nanoparticle Pd-BINAP complexes. The reaction may also be catalyzed by a nanoparticle that consists of two metals.

Organic Redox Reactions

Figure: *Oxidation reaction of cyclohexane to synthesize adiapic acid*

An oxidation reaction to form adipic acid is shown in figure depicted earlier, and it can be catalyzed by cobalt nanoparticles. This is used in an industrial scale to produce the nylon 6,6 polymer. Other examples of oxidation reactions that are catalyzed by metallic nanoparticles include the oxidation of cyclooctane, the oxidation of ethene, and glucose oxidation.

C-C Coupling Reactions

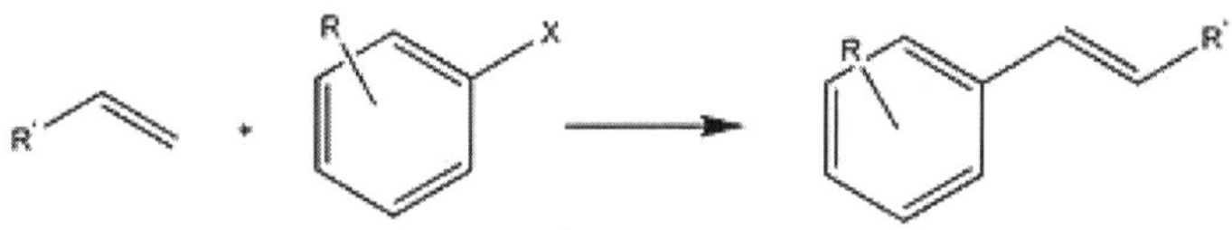

Figure: *Heck coupling reaction*

Metallic nanoparticles can catalyze C–C coupling reactions such as the hydroformylation of olefins, the synthesis of vitamin E and the Heck coupling and Suzuki coupling reactions.

Palladium nanoparticles were found to efficiently catalyze heck coupling reactions. It was found that increased electronegativity of the ligands on the palladium nanoparticles increased their catalytic activity.

The compound $Pd_2(dba)_3$ is a source of Pd(0), which is the catalytically active source of palladium used for many reactions, including cross coupling reactions. Pd2(dba)3 was thought to be a homogeneous catalytic precursor, but recent articles suggest that palladium nanoparticles are formed, making it a heterogeneous catalytic precursor.

Alternative Fuels

Much research on nanomaterial-based catalysts has to do with maximizing the effectiveness of the catalyst coating in fuel cells. Platinum is currently the most common catalyst for this application, however, it is expensive and rare, so a lot of research has been going into maximizing the catalytic properties of other metals by shrinking them to nanoparticles in the hope that someday they will be an efficient and economic alternative to platinum. Gold nanoparticles also exhibit catalytic properties, despite the fact that bulk gold is unreactive.

Fuel cells take advantage of the reaction between hydrogen and oxygen but catalysts are needed to facilitate this reaction. Nanomaterial catalysts can be used to improve energy production. In one experiment, yttrium stabilized zirconium nanoparticles were found to increase the efficiency and reliability of a solid oxide fuel cell.

Nanomaterial ruthenium/platinum catalysts could potentially be used to catalyze the purification of hydrogen for hydrogen storage. Palladium nanoparticles can be functionalized with organometallic ligands to catalyze the oxidation of CO and NO to control air pollution in the environment. Carbon nanotube supported catalysts can be used as a cathode catalytic support for fuel cells and metal nanoparticles have been used to catalyze the growth of carbon nanotubes. Platinum-cobalt bimetallic nanoparticles combined with carbon nanotubes are promising candidates for direct methanol fuel cells since they produce a higher stable current electrode.

Platinum Nanoparticles

Platinum nanoparticles are usually in the form of a suspension or colloid of submicrometre-size particles of platinum in a fluid, usually water. A colloid is technically defined as a stable dispersion of particles in a fluid medium (liquid or gas).

Platinum nanoparticles can be made with sizes between about 2 and 20 nanometres (nm), depending on reaction conditions. Trillions of platinum nanoparticles are suspended in the colloidal solution of brownish-red or black colour. Nanoparticles come in wide variety of shapes including spheres, rods, cubes, and tetrahedra.

Due to the antioxidant properties of the platinum nanoparticles, they are the subject of substantial research, with potential applications in a wide variety of areas, including nanotechnology, catalysis, medicine, and the synthesis of novel materials with unique properties.

Synthesis

Platinum nanoparticles are made by reduction of hexachloroplatinate. A solution is rapidly stirred while a reducing agent such as hydrogen gas is added, causing platinum ions to be reduced to neutral platinum atoms. As more and more of these platinum atoms form, the solution quickly becomes supersaturated and platinum starts to precipitate in the form of sub-nanometre particles. As the reaction proceeds, platinum atoms are added to these nuclei, which grow in size. If the solution is stirred vigorously enough, the particles will be fairly uniform in size and shape. To prevent the particles from aggregating, some sort of stabilizing agent or stabilizer that sticks to the nanoparticle surface is usually added, such as sodium polyacrylate. They can be functionalized with various organic ligands to create organic-inorganic hybrid materials with advanced functionality.

Biological Effects

Research by Yusei Miyamoto at University of Tokyo, Japan, resulted in the use of platinum nanoparticles of the size 2-3 nm to increase the lifespan of the roundworm *Caenorhabditis elegans.*

Nanoparticles may present possible safety issues both medically and environmentally. Most of these issues usually arise due to the high surface-to-volume ratio, which can make the particles of some metals very reactive or catalytic. In particular, inhaled nanoparticles can pose health risks, and may cause inflammation and disease in the lung. They are able to pass through cell membranes in organisms and their interactions with biological systems are relatively unknown.

Fuel and Nanotechnology

How can Nanotechnology Improve Fuel Availability?

Nanotechnology can address the shortage of fossil fuels such as diesel and gasoline by:

- Making the production of fuels from low grade raw materials economical
- Increasing the mileage of engines
- Making the production of fuels from normal raw materials more efficient

Nanotechnology can do all this by increasing the effectiveness of catalysts. Catalysts can reduce the temperature required to convert raw materials into fuel or increase the percentage of fuel burned at a given temperature. Catalysts made from nanoparticles have a greater surface area to interact with the reacting chemicals than catalysts made from larger particles. The larger surface area allows more chemicals to interact with the catalyst simultaneously, which makes the catalyst more effective. This increased effectiveness can make a process such as the production of diesel fuel from coal more economical, and enable the production of fuel from currently unusable raw materials such as low grade crude oil.

Nanotechnology, in the form of genetic engineering, can also improve the performance of enzymes used in the conversion of cellulose into ethanol. Currently ethanol added to gasoline in the United States is made from corn, which is driving up the price of corn. The plan is to use engineered enzymes to break down cellulose into sugar, is fermented to turn the sugar into ethanol. This will allow material that often goes to waste, such as wood chips and grass to be turned into ethanol.

Fuel: Nanotechnology Applications under Development

Reducing the cost of converting crude from oil sands to fuel

Increasing mileage of diesel engines

Nanosphere based catalyst that reduces the cost of producing biodiesel

Modifying crops to allow cellulous material, such as corn stalks to produce enzymes that are triggered at elevated temperatures to convert the cellulous to sugar, simplify the production of ethanol.

Modifying bacteria to cause the production of enzymes that will convert cellulous material to ethanol in one step, rather than converting cellulous to sugar which is than fermented into ethanol.

Tungsten oxide nanoparticles on a material called zirconia used as a nanocatalyst to make the process of refining gasoline more efficient.

Supercapacitor

Supercapacitor (SC), formerly electric double-layer capacitor (EDLC) or ultracapacitor, is the generic term for a family of electrochemical capacitors. Supercapacitors don't have a conventional solid dielectric. The capacitance value of an electrochemical capacitor is determined by two storage principles, which both contribute indivisibly to the total capacitance:

- Double-layer capacitance – Electrostatic storage achieved by separation of charge in a Helmholtz double layer at the interface between the surface of a conductive electrode and an electrolyte. The separation of charge is of the order of a few ångströms (0.3–0.8 nm), much smaller than in a conventional capacitor.
- Pseudocapacitance – Faradaic electrochemical storage with electron charge-transfer, achieved by redox reactions, intercalation or electrosorption.

The ratio of the two storage principles can vary greatly, depending on the design of the electrodes and the composition of the electrolyte.

Pseudocapacitance can be as much as 10× that of double-layer capacitance.

Supercapacitors are divided into three families, based on electrode design:

- Double-layer capacitors – with carbon electrodes or derivates with much higher electrostatic double-layer capacitance than electrochemical pseudocapacitance
- Pseudocapacitors – with metal oxide or conducting polymer electrodes with a high amount of electrochemical pseudocapacitance
- Hybrid capacitors – capacitors with asymmetric electrodes, one of which exhibits mostly electrostatic and the other mostly electrochemical capacitance, such as lithium-ion capacitors

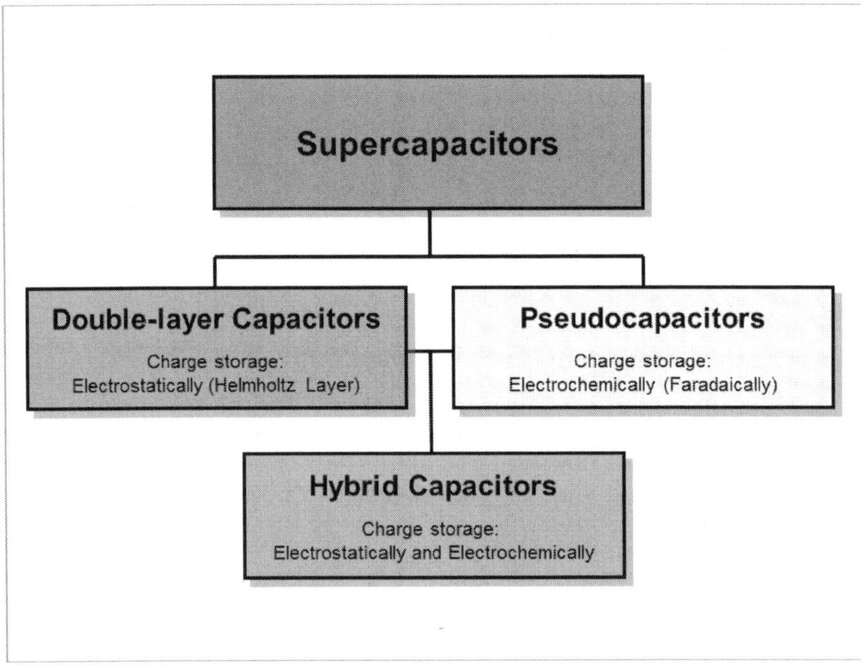

Figure: *Hierarchical classification of supercapacitors and related types*

Supercapacitors bridge the gap between conventional capacitors and rechargeable batteries. They store the most energy per unit volume or mass (energy density) among capacitors. They support up to 10,000 farads/1.2 volt, up to 10,000 times that of electrolytic capacitors, but deliver or accept less than half as much power per unit time (power density).

By contrast, while supercapacitors have energy densities that are approximately 10% of conventional batteries, their power density is generally 10 to 100 times greater. This results in much shorter charge/discharge cycles than batteries. Additionally, they will tolerate many more charge and discharge cycles than batteries.

In these electrochemical capacitors, the electrolyte is the conductive connection between the two electrodes. This distinguishes them from electrolytic capacitors, in which the electrolyte is the cathode and thus forms the second electrode.

Supercapacitors are polarized and must operate with the correct polarity. Polarity is controlled by design with asymmetric electrodes, or, for symmetric electrodes, by a potential applied during manufacture.

Figure: *Maxwell Technologies supercapacitor products*

Supercapacitors support a broad spectrum of applications, including:

- Low supply current for memory backup in static random-access memory (SRAM)
- Power for cars, buses, trains, cranes and elevators, including energy recovery from braking, short-term energy storage and burst-mode power delivery

Exceptional are the manifold different trade or series names used for supercapacitors, such as: APowerCap, BestCap, BoostCap, CAP-XX, DLCAP, EneCapTen, EVerCAP, DynaCap, Faradcap, GreenCap, Goldcap, HY-CAP, Kapton capacitor, Super capacitor, SuperCap, PAS Capacitor, PowerStor, PseudoCap, Ultracapacitor.

Evolution of Components

In the early 1950s, General Electric engineers began experimenting with components using porous carbon electrodes for fuel cells and rechargeable batteries. Activated charcoal is an electrical conductor that is an extremely porous "spongy" form of carbon with a high specific surface area. In 1957 H. Becker developed a "Low voltage electrolytic capacitor with porous carbon electrodes".

He believed that the energy was stored as a charge in the carbon pores as in the pores of the etched foils of electrolytic capacitors. Because the double layer mechanism was not known at the time, he wrote in the patent: "It is not known exactly what is taking place in the component if it is used for energy storage, but it leads to an extremely high capacity."

General Electric did not immediately pursue this work. In 1966 researchers at Standard Oil of Ohio (SOHIO) developed another version of the component as "electrical energy storage apparatus", while working on experimental fuel cell designs. The nature of electrochemical energy storage was not described in this patent. Even in 1970, the electrochemical capacitor patented by Donald L. Boos was registered as an electrolytic capacitor with activated carbon electrodes.

Early electrochemical capacitors used two aluminum foils covered with activated carbon - the electrodes - which were soaked in an electrolyte and separated by a thin porous insulator. This design gave a capacitor with a capacitance value in the one farad area, significantly higher than electrolytic capacitors of the same dimensions. This basic mechanical design remains the basis of most electrochemical capacitors.

SOHIO did not commercialize their invention, licensing the technology to NEC, who finally marketed the results as "supercapacitors" in 1971, to provide backup power for computer memory.

Between 1975 and 1980 Brian Evans Conway conducted extensive fundamental and development work on ruthenium oxide electrochemical capacitors. In 1991 he described the difference between 'Supercapacitor' and 'Battery' behaviour in electrochemical energy storage. In 1999 he coined the term supercapacitor to explain the increased capacitance by surface redox reactions with faradaic charge transfer between electrodes and ions.

His "supercapacitor" stored electrical charge partially in the Helmholtz double-layer and partially as the result of faradaic reactions with "pseudocapacitance" charge transfer of electrons and protons between electrode and electrolyte. The working mechanisms of pseudocapacitors are redox reactions, intercalation and electrosorption. With his research Conway extend the knowledge about electrochemical capacitors up to a new level.

The market expanded slowly. That changed around 1978 as Panasonic marketed its "Goldcaps" brand. This product became a successful energy source for memory backup applications. Competition started only years later. In 1987 ELNA "Dynacap"s entered the market. First generation EDLC's had relatively high internal resistance that limited the discharge current. They were used for low current applications such as powering SRAM chips or for data backup.

At the end of the 1980s improved electrode materials increased capacitance values. At the same time the development of electrolytes with better conductivity lowered the Equivalent Series Resistance (ESR) increasing charge/discharge currents. The first supercapacitor with low internal resistance was developed in 1982 for military applications through the Pinnacle Research Institute (PRI), and were marketed under the brand name "PRI Ultracapacitor". In 1992, Maxwell Laboratories, (later Maxwell Technologies) took over this development. Maxwell adopted the term Ultracapacitor from PRI and called them "Boost Caps" to underline their use for power applications.

Since capacitors' energy content increases with the square of the voltage, researchers were looking for a way to increase the electrolyte's breakdown voltage. In 1994 using the anode of a 200V high voltage tantalum electrolytic capacitor, David A. Evans developed an "Electrolytic-Hybrid Electrochemical Capacitor".

These capacitors combine features of electrolytic and electrochemical capacitors. They combine the high dielectric strength of an anode from an electrolytic capacitor with the high capacitance of a pseudocapacitive metal oxide (ruthenium (IV) oxide) cathode from an electrochemical capacitor, yielding a hybrid electrochemical capacitor. Evans' capacitors, coined Capattery, had an energy content about a factor of 5 higher than a comparable tantalum electrolytic capacitor of the same size. Their high costs limited them to specific military applications.

Recent developments include lithium-ion capacitors. These hybrid capacitors were pioneered by FDK in 2007. They combine an electrostatic carbon electrode with a pre-doped lithium-ion electrochemical electrode. This combination increases the capacitance value. Additionally, the pre-doping process lowers the anode potential and results in a high cell output voltage, further increasing energy density.

Research departments are active in many companies and universities are working to improve characteristics, such as energy density, power density, cycle stability and reduce production costs.

Basics

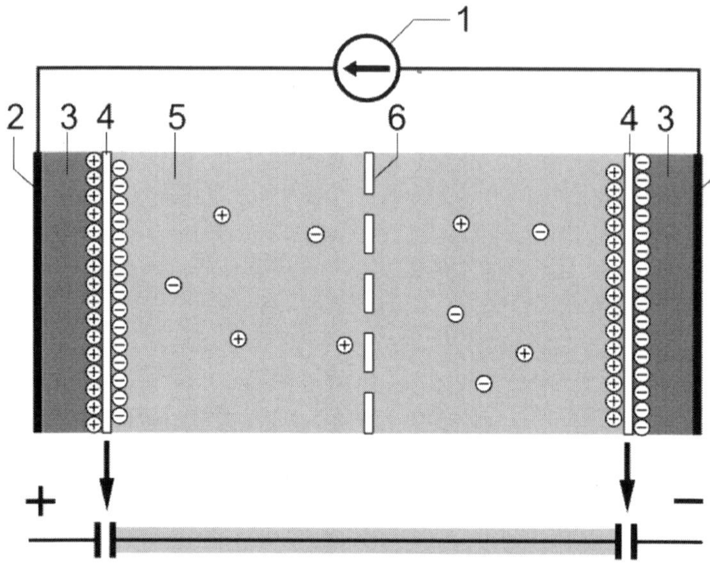

Figure: *Principle construction of a supercapacitor; 1. power source, 2. collector, 3.polarized electrode, 4. Helmholtz double layer, 5. electrolyte having positive and negative ions, 6. Separator.*

Basic Design: Electrochemical capacitors (supercapacitors) consist of two electrodes separated by an ion permeable membrane (separator), and an electrolyte connecting electrically the both electrodes. By applying a voltage to the capacitor an electric double layer at both electrodes is formed, which has a positive or negative layer of ions deposited in a mirror image on the opposite electrode.

Capacitance Distribution

The two electrodes form a series circuit of two individual capacitors C_1 and C_2. The total capacitance C_{total} is given by the formula

$$C_{total} = \frac{C_1 \cdot C_2}{C_1 + C_2}$$

Supercapacitors may have either symmetric or asymmetric electrodes. Symmetry implies that both electrodes have the same capacitance value. If $C_1 = C_2$, then $C_{total} = 0.5 . C_1$. For symmetric capacitors the total capacitance value equals half the value of a single electrode.

For asymmetric capacitors one of the electrodes typically has a higher capacitance value than the other. If $C_1 >> C_2$, then $C_{total} \approx C_2$. Thus, with asymmetric electrodes the total capacitance may be approximately equal to the smaller electrode.

Storage Principles

Electrochemical capacitors uses the double-layer effect to store electric energy. This double-layer has no conventional solid dielectric which separates the charges. The capacitance values of electrochemical capacitors are determined by two new and different high-capacity storage principles in the electric double-layer on their electrodes:

- Double-layer capacitance, electrostatic storage of the electrical energy achieved by separation of charge in a Helmholtz double layer.
- Pseudocapacitance, electrochemical storage of the electrical energy achieved by faradiac redox reactions with charge-transfer.

The amount of charge stored per unit voltage in an electrochemical capacitor is primarily a function of the electrode size but the amount of capacitance of each storage principle can vary extremely. Double-layer capacitance and pseudocapacitance both contribute to the total capacitance value of an electrochemical capacitor. Both capacitances are only separable by measurement techniques.

Practically, these storage principles yield a capacitor with a capacitance value on the order of units to hundreds Farad.

Electrostatic Double-Layer Capacitance

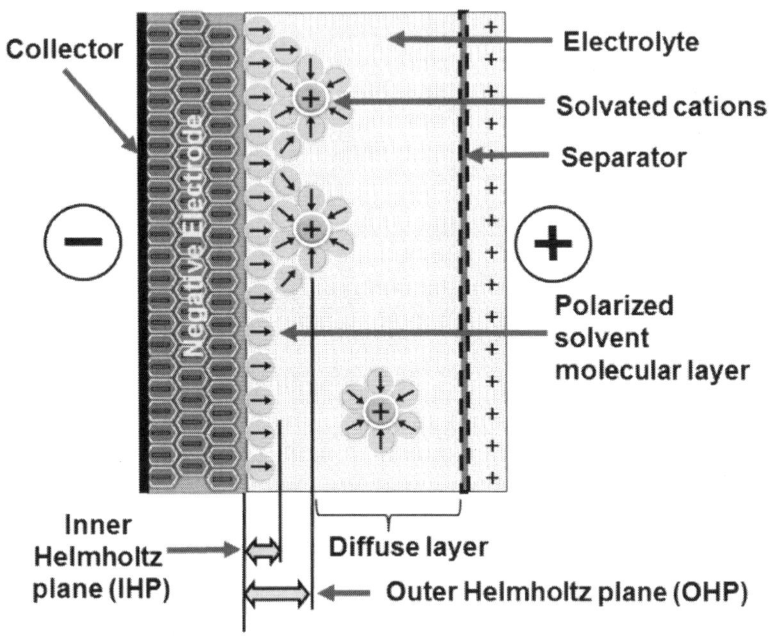

Figure: *Simplified view of a double-layer of negative ions in the electrode and solvated positive ions in the liquid electrolyte, separated by a layer of polarized solvent molecules.*

Every electrochemical capacitor has two electrodes, mechanically separated by a separator, which are electrically connected to each other via the electrolyte. The electrolyte is a mixture of positive and negative ions dissolved in a solvent such as water. At each of the two electrodes surfaces originates an area in which the liquid electrolyte contacts the conductive metallic surface of the electrode. This interface forming a common boundary among two different phases of matter, such as an insoluble solid electrode surface and an adjacent liquid electrolyte. In this interface occurs a very special phenomenon of the double layer effect.

Applying a voltage to an electro-chemical capacitor both electrodes in the capacitor generates electrical double-layers. These double layers consist out of two layers of ions. One layer is in the surface lattice structure of the electrode. The other layer, with opposite polarity, emerges from dissolved and solvated ions in the electrolyte. The two layers are separated by a monolayer of solvent molecules, e. g. for water as solvent by water molecules. The monolayer forms the inner Helmholtz plane (IHP). It adheres by physical adsorption on the surface of the electrode and separates the oppositely polarized ions from each other, becoming a molecular dielectric. The forces that cause the adhesion are not chemical bonds but physical forces. Chemical bonds persist within of the adsorbed molecules, but they are polarized.

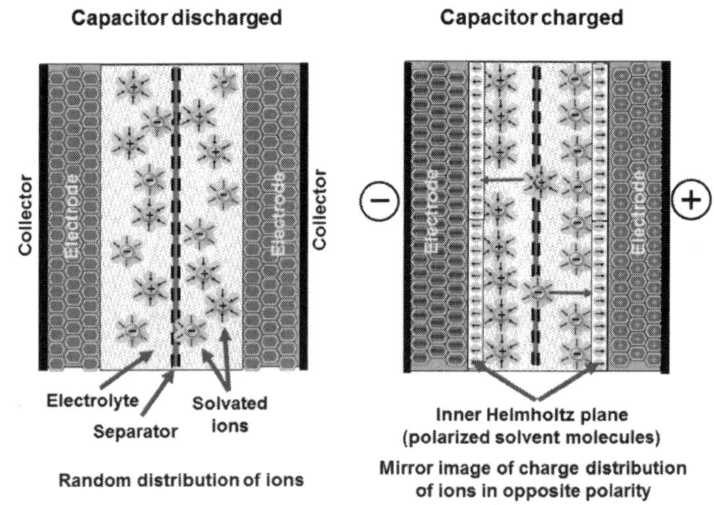

Figure: *Structure and function of an ideal double-layer capacitor. Applying a voltage to the capacitor at both electrodes a Helmholtz double-layer will be formed separating the ions in the electrolyte in a mirror charge distribution of opposite polarity*

The amount of charge in the electrode is matched by the magnitude of counter-charges in outer Helmholtz plane (OHP). This double-layer phenomena store electrical charges as in a conventional capacitor. The double-layer charge forms a static electric field in the molecular layer of the solvent molecules in the IHP that corresponds to the strength of the applied voltage.

The double-layer serves approximately like the dielectric layer in a conventional capacitor, but with the thickness of a single molecule. Therefore to calculate the capacitance the standard formula for conventional plate capacitors can be used. This capacitance can be calculated with:

$$C = \frac{\varepsilon A}{d}.$$

The capacitance C is greatest in capacitors made from materials with a high permittivity ε, large electrode plate surface areas A and reciprocal to the distance d between plates.

If the electrolyte solvent is water then the influence of the high field strength creates permittivity ε of 6 (instead of 80 without an applied electric field). Because activated carbon electrodes have an extremely large surface area in the range of 10 to 40 $\mu F/cm^2$ and the extremely thin double-layer distance is on the order of a few ångströms (0.3-0.8 nm), the double-layer capacitors have much higher capacitance values than conventional capacitors.

The amount of charge stored per unit voltage in an electrochemical capacitor is primarily a function of the electrode size. The electrostatic storage of energy in the double-layers is linear with respect to the stored charge, and correspond to the concentration of the adsorbed ions. But deviating from conventional capacitors, where usually the charge is transferred via electrons, the capacitance of the double-layer capacitors depends from the limited moving speed of ions in the electrolyte and the resistive porous structure of the electrodes.

Capacitance values of supercapacitors depends strongly on the measuring time. Charging and discharging electric double-layers in principle is unlimited. No chemical changes take place. Lifetimes of real supercapacitors only are limited by electrolyte evaporation effects.

Electrochemical Pseudocapacitance

Applying a voltage at the electrochemical capacitor terminals moves the polarized ions or charged atoms in the electrolyte to the opposite polarized electrode and forms a double-layer, separated by a single layer of solvent molecules. Pseudocapacitance can originate when specifically adsorbed cations out of the electrolyte pervade the double-layer. This pseudocapacitance stores electrical energy by means of reversible faradaic redox reactions on the surface of suitable electrodes in an electrochemical capacitor with a electric double-layer.

Pseudocapacitance with specifically adsorbed ions

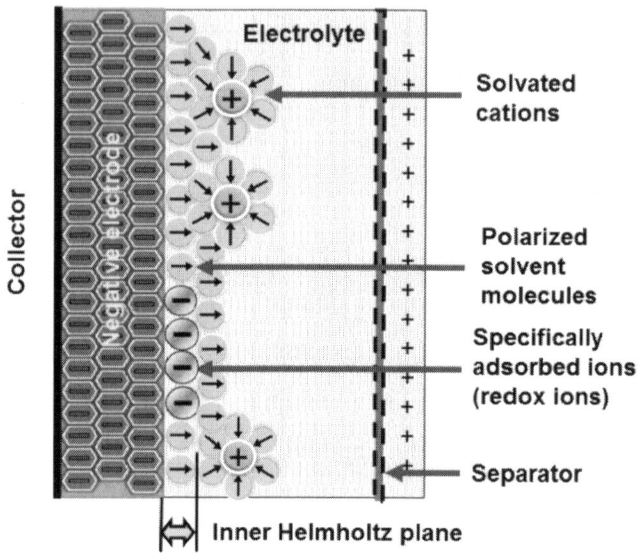

Figure: *Simplified view of a double-layer with specifically adsorbed ions which have submitted their charge to the electrode to explain the faradaic charge-transfer of the pseudocapacitance.*

Pseudocapacitance is accompanied with an electron charge-transfer between electrolyte and electrode coming from a de-solvated and adsorbed ion whereby only one electron per charge unit is participating. This faradaic charge transfer originates by a very fast sequence of reversible redox, intercalation or electrosorption processes. The adsorbed ion has no chemical reaction with the atoms of the electrode. No chemical bonds arise. Only a charge-transfer take place.

The electrons involved in the faradaic processes are transferred to or from valence electron states (orbitals) of the redox electrode reagent. They enter the negative electrode and flow through the external circuit to the positive electrode where a second double-layer with an equal number of anions has formed. But these anions don't accept the electrons. They remain on the electrode's surface in the charged state, and the electrons remain in the strongly ionized and "electron hungry" transition-metal ions of the electrode. This kind of pseudocapacitance has a linear function within narrow limits and is determined by the potential-dependent degree of surface coverage of the adsorbed anions. The storage capacity of the pseudocapacitance is limited by the finite quantity of reagent of available surface. A faradaic pseudocapacitance still only occurs together with a static double-layer capacitance. Pseudocapacitance and double-layer capacitance both contribute indivisible to the total capacitance value of the chemical capacitor. The amount of pseudocapacitance depends on the surface area, material and structure of the electrodes.

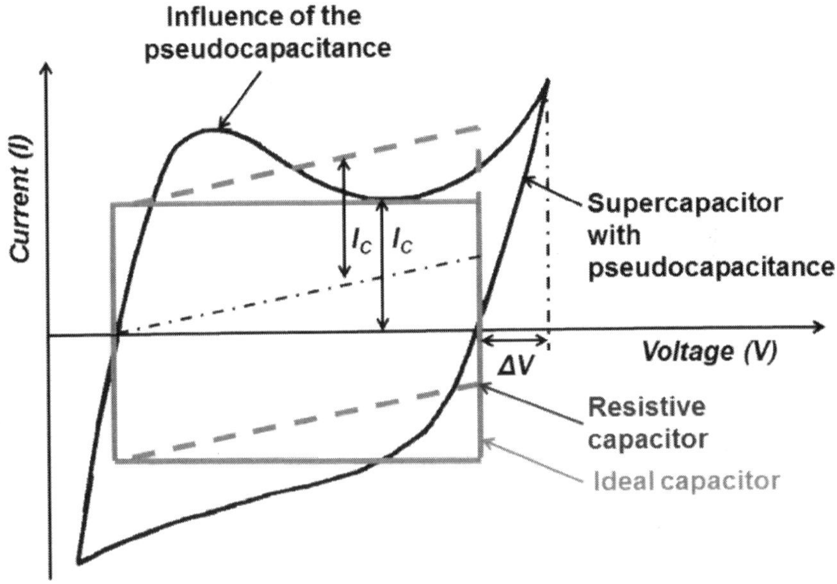

Figure: *A cyclic voltammogram shows the fundamental differences between static capacitance (rectangular) and pseudocapacitance (curved)*

Since all the pseudocapacitance reactions take place only with de-solvated ions, which are much smaller than solvated ion with their solvating shell, the pseudocapacitance can be much higher than the double-layer capacitance for the same electrode surface. Therefore the pseudocapacitance may exceed the value of double-layer capacitance for the same surface area by factor 100, depending on the nature and the structure of the electrode.

The ability of electrodes to accomplish pseudocapacitance effects by redox reactions, intercalation or electrosorption strongly depends on the chemical affinity of electrode materials to the ions adsorbed on the electrode surface as well as on the structure and dimension of the electrode pores. Materials exhibiting redox behaviour for use as electrodes in pseudocapacitors are transition-metal oxides like ruthenium (RuO_2), iridium (IrO_2), or manganese (MnO_2) inserted by doping in the conductive electrode material such as active carbon, as well as conducting polymers such as polyaniline or derivatives of polythiophene covering the electrode material.

Pseudocapacitance may also originate from the structure and especially from the pore size of the electrodes. The tailored sizes of pores in nano-structured carbon electrodes like carbide-derived carbons (CDCs) or carbon nanotubes (CNTs) for electrodes can be referred to as intercalated pores which can be entered by de-solvated ions from the electrolyte solution and originates pseudocapacitance. The amount of electric charge stored in a pseudocapacitance is linearly proportional to the applied voltage. The unit of pseudocapacitance is farad.

Potential Distribution

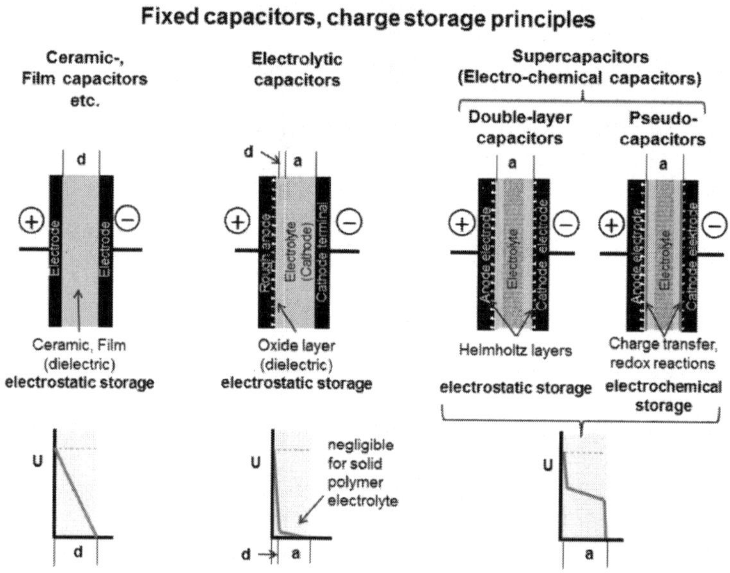

Figure: Charge storage principles of different capacitor types and their internal potential distribution

Conventional capacitors consist out of two electrodes which are separated by a dielectric material. In conventional capacitors such as ceramic capacitors and film capacitors, the electric charge of a loaded capacitor is stored in a static electric field that permeates the dielectric between the electrodes. The electric field originates by the separation of charge carriers and the strength of the electric field correlates with the potential between the two electrodes. It drops over the dielectric. The total energy increases with the amount of stored charge and the potential between the plates. The maximum potential between the plates, the maximal voltage, is limited by the dielectric's breakdown field strength.

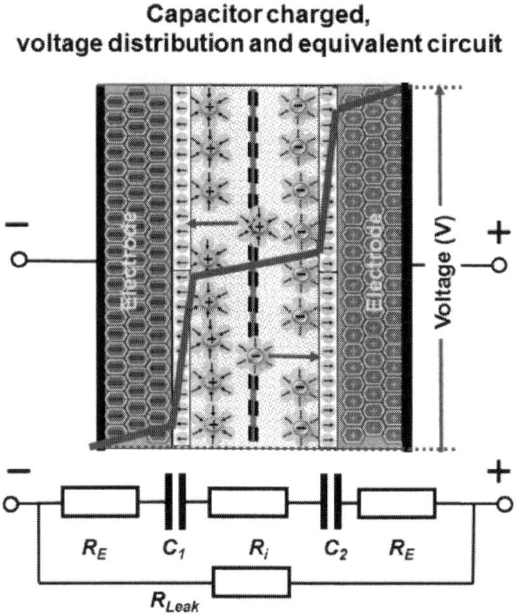

Figure: *Basic illustration of the functionality of a supercapacitor, the voltage distribution inside of the capacitor and its simplified equivalent DC circuit*

This static storage also applies for electrolytic capacitors in which most of the potential decreases over the anode's thin oxide layer. The electrolyte as cathode may be somewhat resistive so that for "wet" electrolytic capacitors, a small amount of the potential decreases over the electrolyte. For electrolytic capacitors with solid conductive polymer electrolyte this voltage drop is negligible.

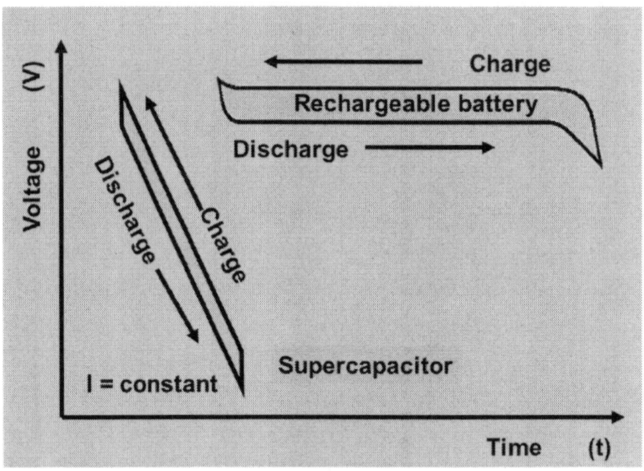

Figure: *The voltage behaviour of supercapacitors and batteries during charging/discharging differs clearly*

Conventional capacitors are also called electrostatic capacitors. The potential (voltage) of a charged capacitor correlates linearly with the stored charge.

Different from conventional capacitors electrochemical capacitors (supercapacitors) basically consists out of two electrodes separated by an ion permeable membrane (separator), and electrically connected via an electrolyte. In this double-layer electrodes a mixture of a double-layer and pseudocapacitance is stored. If both electrodes have approximately the same resistance (#Internal resistance), the potential of the capacitor decreases symmetrically over both double-layers, whereby a voltage drop across the ESR of the electrolyte is achieved. The maximum potential across the capacitor, the maximal voltage, is limited by the electrolytes decomposition voltage.

Both the electrostatic and electrochemical storage of energy in electrochemical capacitors are linear with respect to the stored charge, just as in conventional capacitors. The voltage between the capacitors terminals is linear with respect to the amount of stored energy. This linear voltage gradient differs from rechargeable electrochemical batteries, in which the voltage between the terminals remains independent of the amount of stored energy, providing a relatively constant voltage.

Construction

Construction Details:

- Styles of supercapacitors with activated carbon electrodes

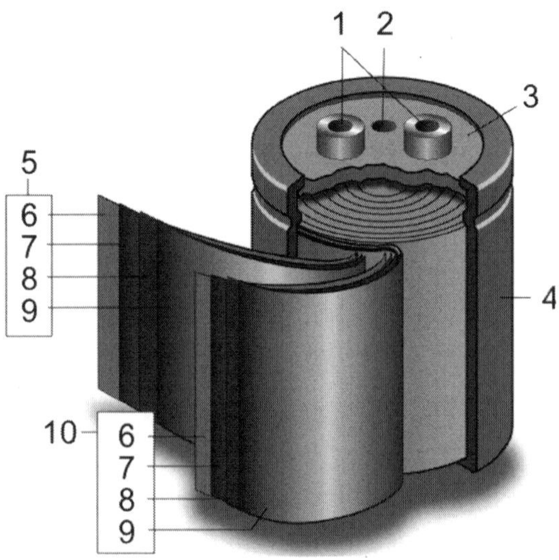

Figure: *Schematic construction of a wound supercapacitors*

1.Terminals, 2.Safety vent, 3.Sealing disc, 4.Aluminum can, 5.Positive pole, 6.Separator, 7.Carbon electrode, 8.Collector, 9.Carbon electrode, 10.Negative pole

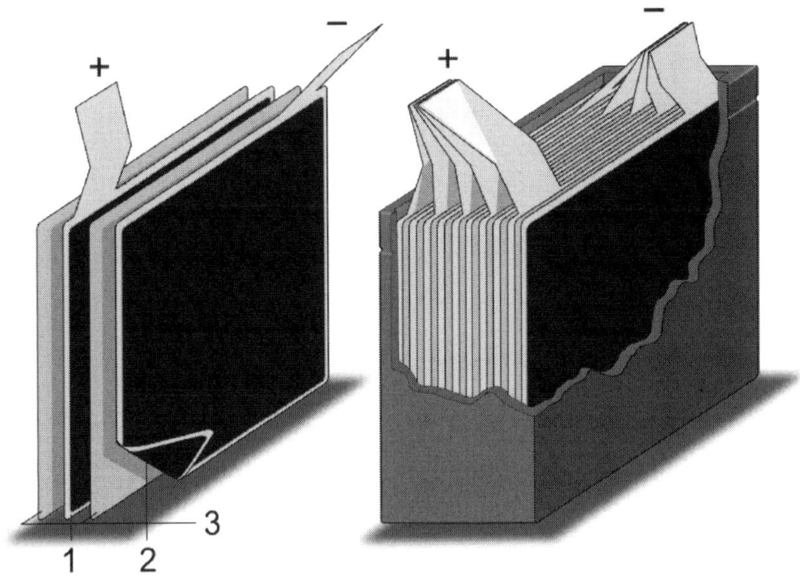

Figure: *Schematic construction of a supercapacitor with stacked electrodes 1.Positive electrode, 2.Negative electrode, 3.Separator*

Supercapacitors are constructed with two metal foils (current collectors), each coated with an electrode material e.g. activated carbon. The collectors serves as the power connection between the electrode material and the external terminals of the capacitor.

Specifically of the electrode material is the structure having a very large surface. E. g. in this example activated carbon is electrochemically etched, so that the surface of the material is about a factor 100,000 larger than the smooth surface.

The both electrodes are separated by an ion permeable membrane (separator) used as insulator to protect the electrodes against direct contact forcing short circuits. This construction is subsequently rolled or folded into a cylindrical or rectangular shape and can be stacked in an aluminum can or an adaptable rectangular housing.

Then the cell is impregnated with a liquid or viscous electrolyte organic or aqueous type, or may be of solid state. The electrolyte, an ionic conductor enters the pores of the electrodes and serves as conductive connection between the electrodes across the separator. Last but not least the housing will be hermetically closed to ensure stable behaviour over the specified life time.

Materials

The properties of supercapacitors come from the interaction of their internal materials. Especially the combination of electrode material and kind of electrolyte determine the functionality and the thermic and electrical characteristics of the capacitors.

Additional the three members of the supercapacitor family are determined by their electrodes material and structure.

Supercapacitor Types

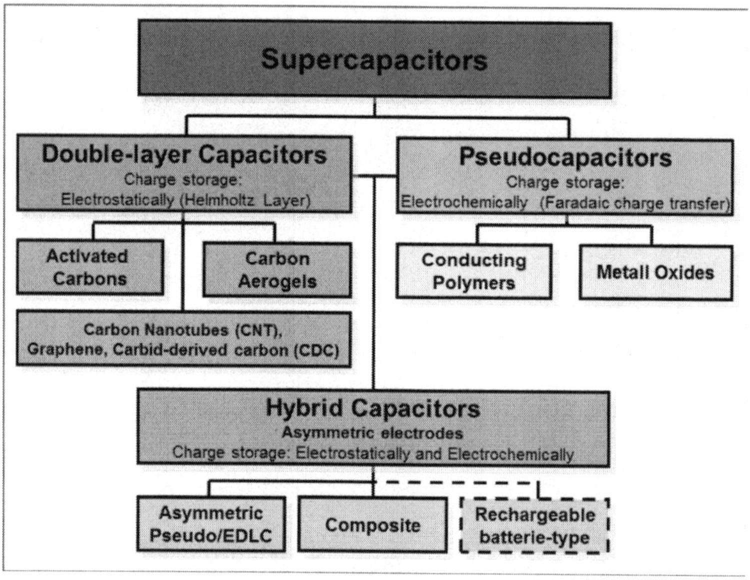

Figure: *Family tree of supercapacitor types. Double-layer capacitors and pseudocapacitors as well as hybrid capacitors are defined over their electrode designs.*

Supercapacitors store, as described above, its electric energy with the two different storage principles, the static double-layer capacitance and electrochemical pseudocapacitance. The distribution of the amounts of both capacitances per capacitor depends on the material and structure of the electrodes. Based on this the supercapacitor family are divided into three types:

- Double-layer capacitors – with activated carbon electrodes or derivates with much higher electrostatic double-layer capacitance than electrochemical pseudocapacitance
- Pseudocapacitors – with transition metal oxide or conducting polymer electrodes with a high amount of electrochemical pseudocapacitance

- Hybrid capacitors – capacitors with asymmetric electrodes one of which exhibits mostly electrostatic and the other mostly electrochemical capacitance, such as lithium-ion capacitors

Electrodes

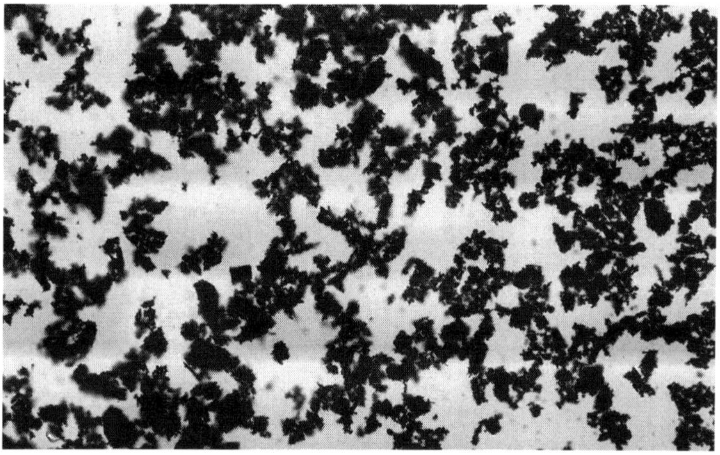

Figure: *A micrograph of activated charcoal under bright field illumination on a light microscope. Notice the fractal-like shape of the particles hinting at their enormous surface area. Each particle in this image, despite being only around 0.1 mm across, has a surface area of several square metres.*

The electrodes for supercapacitors are generally thin coatings applied and electrical connected to a conductive, metallic current collector. Electrodes must exhibit good conductivity, high temperature stability, longtime chemical stability (inert), high corrosion resistance and high surface areas per unit volume and mass. Other requirements include environmental friendliness and low cost.

The amount of double-layer as well as pseudocapacitance stored per unit voltage in a supercapacitor is predominate a function of the electrode surface area. Therefore the electrodes for supercapacitors typically are made out of porous, "spongy" material with an extraordinarily high specific surface area, such as activated carbon. Additionally, the ability of the electrode material to perform faradaic charge transfers enhances the total capacitance.

Structurally, pore sizes in carbons range from micropores (less than 2 nm) to mesopores (2-50 nm) but below macropores (greater than 50 nm). Pseudocapacitance requires micropore that are accessible only to de-solvated ions.

Generally the smaller the electrode's pores, the greater the capacitance and energy density. However, smaller pores increase (ESR) and decrease power density. Applications with high peak currents require larger pores and low internal losses, while applications requiring high energy density need small pores.

Electrodes for EDLCs

The most commonly used electrode material for supercapacitors is carbon in various manifestations such as activated carbon (AC), carbon fibre-cloth (AFC), carbide-derived carbon (CDC), carbon aerogel, graphite (graphene), and carbon nanotubes (CNTs).

Activated Carbon

Activated carbon (AC) was the first material chosen for EDLC electrodes. It has an electrical conductivity of 1,250 to 3,000 S/m, approximately 0.003% of metallic conductivity, but sufficient for supercapacitors.

Activated charcoal is an extremely porous form of carbon with a high specific surface area — a common approximation is that 1 gram (0.035 oz) (a pencil-eraser-sized amount) has a surface area of roughly 1,000 to 3,000 square metres (11,000 to 32,000 sq ft) — about the size of 4 to 12 tennis courts. It is typically a powder made of fine but "rough" particles. The bulk form used in electrodes is low-density with many pores, giving high double-layer capacitance.

Solid activated carbon, also termed *consolidated amorphous carbon* (CAC) is the most used electrode material for supercapacitors and may be cheaper than other carbon derivatives. It is produced from activated carbon powder pressed into the desired shape, forming a block with a wide distribution of pore sizes. An electrode with a surface area of about 1000 m^2/g results in a typical double-layer capacitance of about 10 $\mu F/cm^2$ and a specific capacitance of 100 F/g.

As of 2010 virtually all commercial supercapacitors use powdered activated carbon made from environmentally friendly coconut shells. Coconut shells produce activated carbon with more micropores than with charcoal from wood.

Activated carbon electrodes exhibit predominantly static double-layer capacitance, but also exhibit pseudocapacitance. Pores with diameters <2 nm are accessible only to de-solvated ions and enable faradaic reactions.

Carbon nanotubes offer 3x more surface area than activated carbon. Higher performance, higher cost components are available, based on synthetic carbon precursors that are activated with potassium hydroxide (KOH).

Activated Carbon Fibres

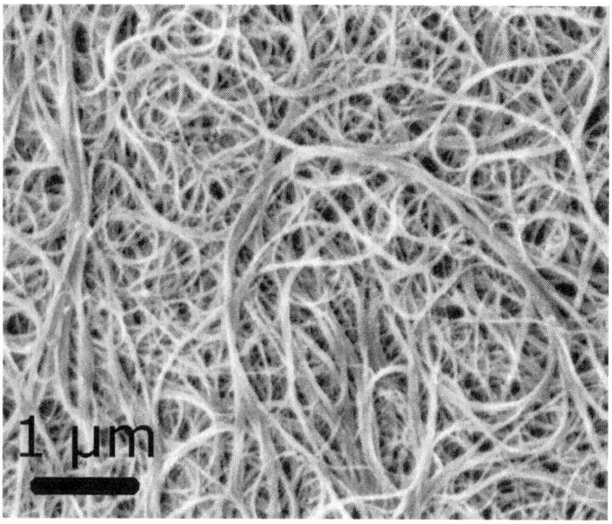

Figure: *SEM image of carbon nanotube bundles with a surface of about 1500 m^2/g*

Activated carbon fibres (ACF) are produced from activated carbon and have a typical diameter of 10 µm. They can have micropores with a very narrow pore-size distribution that can be readily controlled. The surface area of AFC woven into a textile is about 2500 m^2/g. Advantages of AFC electrodes include low electrical resistance along the fibre axis and good contact to the collector.

AFC electrodes exhibit predominantly double-layer capacitance with a small amount of pseudocapacitance due to their micropores.

Carbon Aerogel

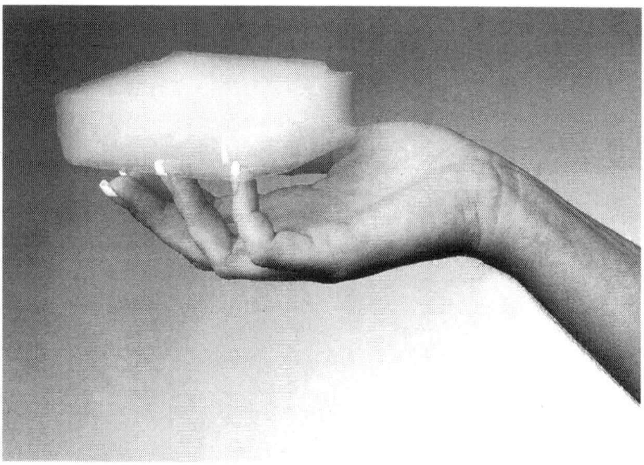

Figure: *A block of aerogel in hand*

Carbon aerogel is a highly porous, synthetic, ultralight material derived from an organic gel in which the liquid component of the gel has been replaced with a gas. It is also called "frozen smoke".

Aerogel electrodes are made via pyrolysis of resorcinol formaldehyde aerogels. Carbon aerogels are more conductive than most activated carbons. They enable thin and mechanically stable electrodes with a thickness in the range of several hundred micrometres (µm) and with uniform pore size. It also exhibits mechanical and vibration stability for supercapacitors in high vibration applications.

Standard aerogel electrodes exhibit predominantly double-layer capacitance. Aerogel electrodes that incorporate composite material can add a high amount of pseudocapacitance.

Researchers have created a carbon aerogel electrode with gravimetric densities of about 400–1200 m^2/g and specific capacitance of 104 F/cm^3, yielding in energy density of 325 J/g (90 W•h/kg) and power density of 20 W/g.

In 2013, an aerographene with a volumetric density of 0.16 mg/cm^3 was synthesized, becoming the lightest known material.

Carbide-Derived Carbon

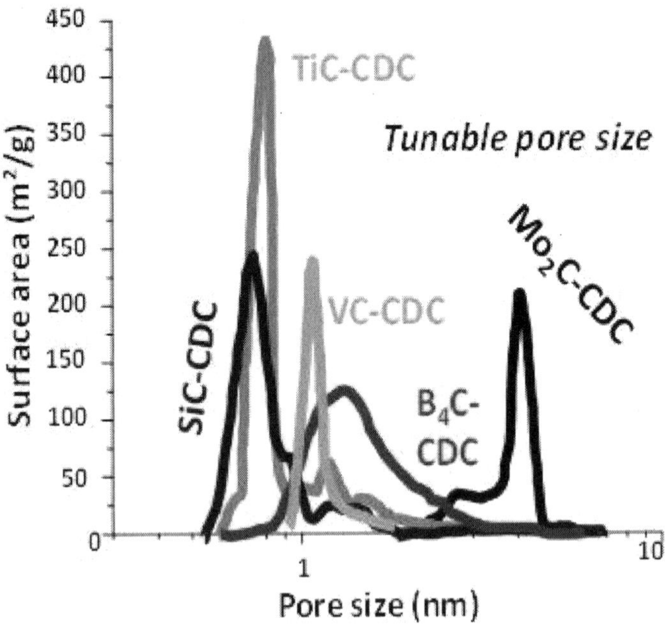

Figure: *Pore size distributions for different carbide precursors.*

Carbide-derived carbon (CDC), also known as tunable nanoporous carbon, is a family of carbon materials derived from carbide precursors, such as binary silicon carbide and titanium carbide, that are transformed into pure carbon via physical (e.g., thermal decomposition) or chemical (e.g., halogenation) processes.

Carbide-derived carbons can exhibit high surface area and tunable pore diameters to maximize ion confinement, increasing pseudocapacitance by faradaic H_2 adsorption treatment. Structurally, CDC pore sizes range from micropores to mesopores. Capacitance may be increased by using micropores. Sub-1 nm pores contribute to capacitance even if the solvated ions are larger. This capacitance increase is explained by the distortion of the ion-solvating shell. As pore size approaches the solvation shell size, solvent molecules are excluded and de-solvated ions fill the pores, increasing ionic packing density and storage capability by faradaic H_2 intercalation. CDC electrodes with tailored pore design offer as much as 75% greater energy density than conventional activated carbons.

In 2013 a CDC supercapacitor offered an energy density of 8.3 Wh/kg having 4,000 F capacitance and one million charge/discharge cycles.

Graphene

Figure: *Graphene is an atomic-scale honeycomb lattice made of carbon atoms.*

Graphene is a one-atom thick sheet of graphite, with atoms arranged in a regular hexagonal pattern, also called "nanocomposite paper".

Graphene has a surface area of 2630 m^2/g which can lead theoretically to a capacitor of 550 F/g. Its important advantage is high conductivity >1700 S/m compared to activated carbon (10 to 100 S/m). As of 2012 a new development used graphene sheets directly as electrodes without collectors for portable applications.

One graphene-based supercapacitor uses curved graphene sheets that do not stack face-to-face, forming mesopores that are accessible to and wettable by environmentally friendly ionic electrolytes at voltages up to 4 V. They have a specific energy density of 85.6 W•h/kg at room temperature equaling that of a conventional nickel metal hydride battery, but with 100-1000 times greater power density.

The two-dimensional structure of graphene improves charging and discharging. Charge carriers in vertically oriented sheets can quickly migrate into or out of the deeper structures of the electrode, thus increasing currents. Such capacitors may be suitable for 100/120 Hz filter applications, which are unreachable for supercapacitors using other carbons.

As of 2013 graphene can be produced in various labs, but is not available in production quantities.

Carbon Nanotubes

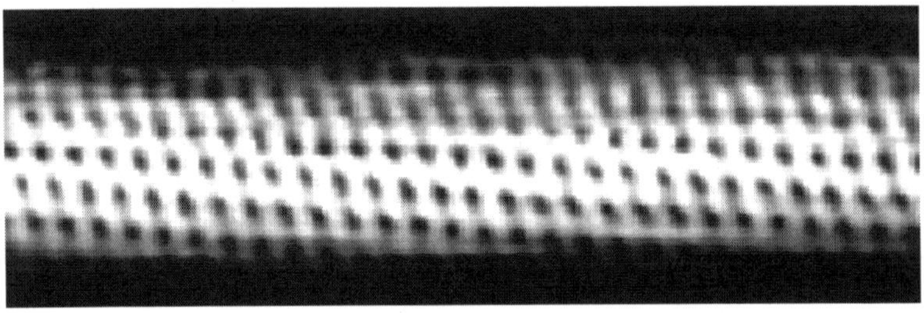

***Figure:** A scanning tunneling microscopy image of single-walled carbon nanotube*

Carbon nanotubes (CNTs), also called buckytubes, are carbon molecules with a cylindrical nanostructure. They have a hollow structure with walls formed by one-atom-thick sheets of graphene. These sheets are rolled at specific and discrete ("chiral") angles, and the combination of chiral angle and radius controls properties such as electrical conductivity, electrolyte wettability and ion access. Nanotubes are categorized as single-walled nanotubes (SWNTs) or multi-walled nanotubes (MWNTs). The latter have one or more outer tubes successively enveloping a SWNT, much like the Russian matroyska dolls. SWNTs have diameters ranging between 1 and 3 nm. MWNTs have thicker coaxial walls, separated by spacing (0.34 nm) that is close to graphene's interlayer distance.

Nanotubes can grow vertically on the collector substrate, such as a silicon wafer. Typical lengths are 20 to 100 μm.

Carbon nanotubes can greatly improve capacitor performance, due to the highly wettable surface area and high conductivity.

CNTs can store about the same charge as activated carbon per unit surface area, but nanotubes' surface is arranged in a regular pattern, providing greater wettability. SWNTs have a high theoretical specific surface area of 1315 m^2/g, while MWNTs' SSA is lower and is determined by the diameter of the tubes and degree of nesting, compared with a surface area of about 3000 m^2/g of activated carbons. Nevertheless, CNTs have higher capacitance than activated carbon electrodes, e.g., 102 F/g for MWNTs and 180 F/g for SWNTs.

MWNTs have mesopores that allow for easy access of ions at the electrode/electrolyte interface. As the pore size approaches the size of the ion solvation shell, the solvent molecules are partially stripped, resulting in larger ionic packing density and increased faradaic storage capability. However, the considerable volume change during repeated intercalation and depletion decreases their mechanical stability. To this end, research to increase surface area, mechanical strength, electrical conductivity and chemical stability is ongoing.

Electrodes for Pseudocapacitors

Pseudocapacitance with faradaic charge transfer also is always present in carbon double-layer electrodes. But the amount of pseudocapacitance in EDLC electrodes is relatively low. Pseudocapacitance electrodes have surfaces able to achieve sufficient faradaic processes to have a predominate amount of pseudocapacitance. Pseudocapacitance electrodes without double-layer capacitance doesn't exist.

Metal Oxides

B. E. Conway's research described electrodes of transition metal oxides that exhibited high amounts of pseudocapacitance. Oxides of transition metals including ruthenium (RuO_2), iridium (IrO_2), iron (Fe_3O_4), manganese (MnO_2) or sulfides such as titanium sulfide (TiS_2) alone or in combination generate strong faradaic electron–transferring reactions combined with low resistance. Ruthenium dioxide in combination with H_2SO_4 electrolyte provides specific capacitance of 720 F/g and a high energy density of 26.7 Wh/kg.

Charge/discharge takes place over a window of about 1.2 V per electrode. This pseudocapacitance of about 720 F/g is roughly 100 times higher than for double-layer capacitance using activated carbon electrodes. These transition metal electrodes offer excellent reversibility, with several hundred-thousand cycles. However, ruthenium is expensive and the 2,4 V voltage window for this capacitor limits their applications to military and space applications.

Less expensive oxides of iron, vanadium, nickel and cobalt have been tested in aqueous electrolytes, but none has been investigated as much as manganese dioxide (MnO2). However, none of these oxides are in commercial use.

Conductive Polymers

Another approach uses electron-conducting polymers as pseudocapacitive material. Although mechanically weak, conductive polymers have high conductivity, resulting in a low ESR and a relatively high capacitance. Such conducting polymers include polyaniline, polythiophene, polypyrrole and polyacetylene. Such electrodes employ also electrochemical doping or dedoping of the polymers with anions and cations. Electrodes out of or coated with conductive polymers are cost comparable to carbon electrodes.

Conducting polymer electrodes generally suffer from limited cycling stability. However, polyacene electrodes provide up to 10,000 cycles, much better than batteries.

Electrodes for Hybrid Capacitors

All commercial hybrid supercapacitors are asymmetric. They combine an electrode with high amount of pseudocapacitance with an electrode with a high amount of double-layer capacitance. In such systems the faradaic pseudocapacitance electrode with their higher capacitance provides high energy density while the non-faradaic EDLC electrode enables high power density. An advantage of the hybride-type supercapacitors compared with symmetrical EDLC's is their higher specific capacitance value as well as their higher rated voltage and correspondingly their higher specific energy.

Composite Electrodes

Composite electrodes for hybrid-type supercapacitors are constructed from carbon-based material with incorporated or deposited pseudocapacitive active materials like metal oxides and conducting polymers. As of 2013 most research for supercapacitors explores composite electrodes.

CNTs give a backbone for a homogeneous distribution of metal oxide or electrically conducting polymers (ECPs), producing good pseudocapacitance and good double-layer capacitance. These electrodes achieve higher capacitances than either pure carbon or pure metal oxide or polymer-based electrodes. This is attributed to the accessibility of the nanotubes' tangled mat structure, which allows a uniform coating of pseudocapacitive materials and three-dimensional charge distribution.

Another way to enhance CNT electrodes is by doping with a pseudocapacitive dopant as in lithium-ion capacitors. In this case the relatively small lithium atoms intercalate between the layers of carbon. The anode is made of lithium-doped carbon, which enables lower negative potential with a cathode made of activated carbon. This results in a larger voltage of 3.8-4 V that prevents electrolyte oxidation. As of 2007 they had achieved capacitance of 550 F/g. and reach an energy density up to 14 Wh/kg.

Battery-Type Electrodes

Rechargeable battery electrodes influenced the development of electrodes for new hybrid-type supercapacitor electrodes as for lithium-ion capacitors. Together with an carbon EDLC electrode in an asymmetric construction offers this configuration higher energy density than typical supercapacitors with higher power density, longer cycle life and faster charging and recharging times than batteries.

While their structure qualifies them as composite electrodes, they are typically placed in the category of composite electrodes.

Asymmetric Electrodes (Pseudo/EDLC)

Recently some asymmetric hybrid supercapacitors were developed in which the positive electrode where based on a real pseudocapacitive metal oxides electrode (not a composite electrode), and the negative electrode on an EDLC activated carbon electrode.

An advantage of this type of supercapacitors is their higher voltage and correspondingly their higher specific energy (up to 10-20 Wh/kg).

As far as known no commercial offered supercapacitors with such kind of asymmetric electrodes are on the market.

Electrolytes

Electrolytes consist of a solvent and dissolved chemicals that dissociate into positive cations and negative anions, making the electrolyte electrically conductive. The more ions the electrolyte contains, the better its conductivity. In supercapacitors electrolytes are the electrically conductive connection between the two electrodes. Additionally, in supercapacitors the electrolyte provides the molecules for the separating monolayer in the Helmholtz double-layer and delivers the ions for pseudocapacitance.

The electrolyte determines the capacitor's characteristics: its operating voltage, temperature range, ESR and capacitance. With the same activated carbon electrode an aqueous electrolyte achieves capacitance values of 160 F/g, while an organic electrolyte achieves only 100 F/g.

The electrolyte must be chemically inert and not chemically attack the other materials in the capacitor to ensure long time stable behaviour of the capacitor's electrical parameters. The electrolyte's viscosity must be low enough to wet the porous, sponge-like structure of the electrodes. An ideal electrolyte does not exist, forcing a compromise between performance and other requirements.

Aqueous

Water is a relatively good solvent for inorganic chemicals. Treated with acids such as sulfuric acid (H_2SO_4), alkalis such as potassium hydroxide (KOH), or salts such as quaternary phosphonium salts, sodium perchlorate ($NaClO_4$), lithium perchlorate ($LiClO_4$) or lithium hexafluoride arsenate (LiAsF6), water offers relatively high conductivity values of about 100 to 1000 mS/cm. Aqueous electrolytes have a dissociation voltage of 1.15 V per electrode (2,3 V capacitor voltage) and a relatively low operating temperature range. They are used in supercapacitors with low energy density and high power density.

Organic

Electrolytes with organic solvents such as acetonitrile, propylene carbonate, tetrahydrofuran, diethyl carbonate, γ-butyrolactone and solutions with quaternary ammonium salts or alkyl ammonium salts such as tetraethylammonium tetrafluoroborate (N(Et)\4BF4) or triethyl (metyl) tetrafluoroborate (NMe(Et)3BF4) are more expensive than aqueous electrolytes, but they have a higher dissociation voltage of typically 1.35 V per electrode (2,7 V capacitor voltage), and a higher temperature range. The lower electrical conductivity of organic solvents (10 to 60 S/cm) leads to a lower power density, but since the energy density increases with the square of the voltage, a higher energy density.

Separators

Separators have to physically separate the two electrodes to prevent a short circuit by direct contact. It can be very thin (a few hundredths of a millimetre) and must be very porous to the conducting ions to minimize ESR. Furthermore, separators must be chemically inert to protect the electrolyte's stability and conductivity. Inexpensive components use open capacitor papers. More sophisticated designs use nonwoven porous polymeric films like polyacrylonitrile or Kapton, woven glass fibres or porous woven ceramic fibres.

Collectors and Housing

Current collectors connect the electrodes to the capacitor's terminals. The collector is either sprayed onto the electrode or is a metal foil. They must be able to distribute peak currents of up to 100 A.

If the housing is made out of a metal (typically aluminum) the collectors should be made from the same material to avoid forming a corrosive galvanic cell.

Electrical Parameters

Capacitance

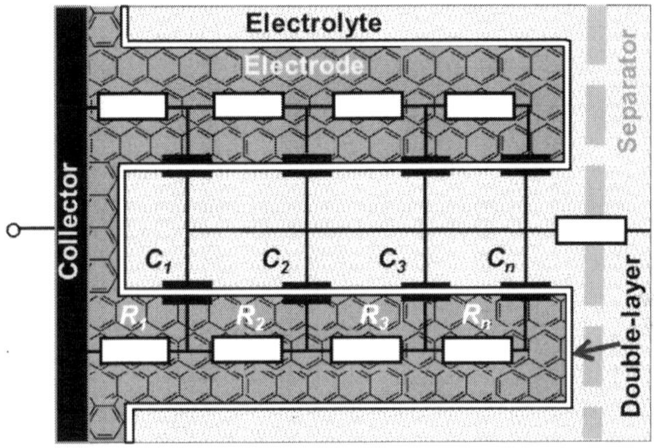

Figure: *Schematic illustration of the capacitance behaviour resulting out of the porous structure of the electrodes*

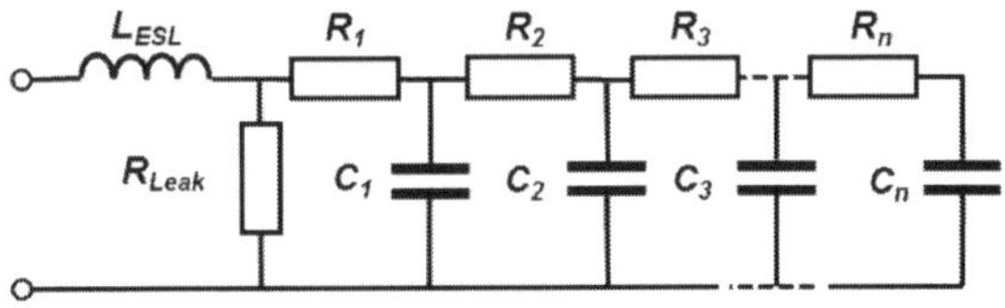

Figure: *Equivalent circuit with cascaded RC elements*

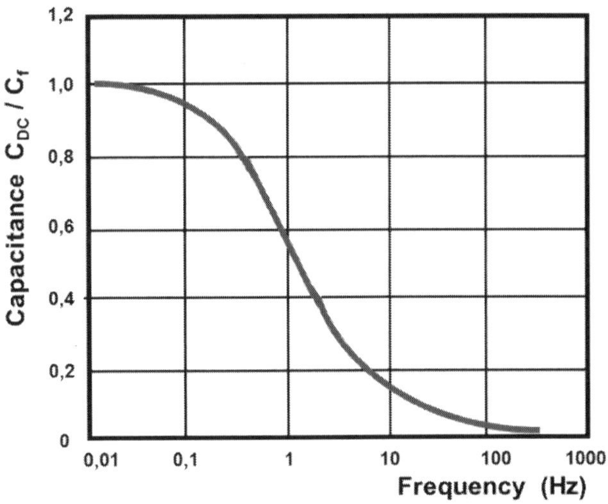

Figure: *Frequency depending of the capacitance value of a 50 F supercapacitor*

Capacitance values for commercial capacitors are specified as "rated capacitance C_R". This is the value for which the capacitor has been designed. The value for an actual component must be within the limits given by the specified tolerance. Typical values are in the range of farads (F), three to six orders of magnitude larger than those of electrolytic capacitors. The capacitance value results from the energy W of a loaded capacitor loaded via a DC voltage V_{DC}.

$$W = \frac{1}{2} \cdot C_{DC} \cdot V_{DC}^2$$

This value is also called the "DC capacitance".

Measurement

Conventional capacitors are normally measured with a small AC voltage (0.5 V) and a frequency of 100 Hz or 1 kHz depending on the capacitor type. The AC capacitance measurement offers fast results, important for industrial production lines. The capacitance value of a supercapacitor depends strongly on the measurement frequency, which is related to the porous electrode structure and the limited electrolyte's ion mobility. Even at a low frequency of 10 Hz, the measured capacitance value drops from 100 to 20 percent of the DC capacitance value.

This extraordinary strong frequency dependence can be explained by the different distances the ions have to move in the electrode's pores. The area at the beginning of the pores can easily be accessed by the ions. The short distance is accompanied by low electrical resistance. The greater the distance the ions have to cover, the higher the resistance. This phenomenon can be described with a series circuit of cascaded RC (resistor/capacitor) elements with serial RC time constants. These result in delayed current flow, reducing the total electrode surface area that can be covered with ions if polarity changes – capacitance decreases with increasing AC frequency. Thus, the total capacitance is only achieved after longer measuring times.

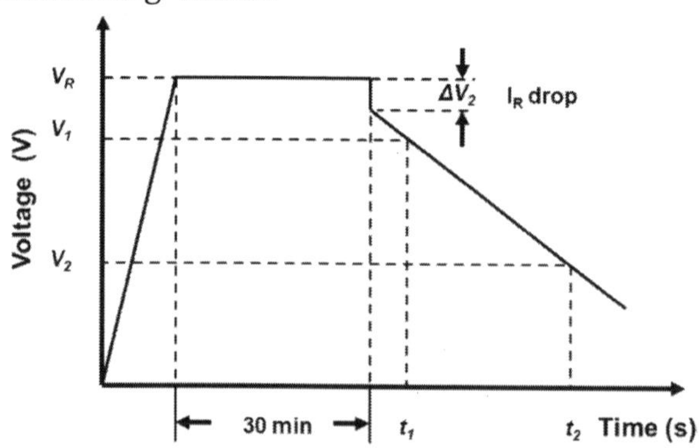

Figure: *Illustration of the measurement conditions for measuring the capacitance of supercapacitors*

Out of the reason of the very strong frequency dependence of the capacitance this electrical parameter has to be measured with a special constant current charge and discharge measurement, defined in IEC standards 62391-1 and -2.

Measurement starts with charging the capacitor. The voltage has to be applied and after the constant current/constant voltage power supply has achieved the rated voltage, the capacitor has to be charged for 30 minutes. Next, the capacitor has to be discharged with a constant discharge current $I_{discharge}$. Than the time t_1 and t_2, for the voltage to drop from 80% (V_1) to 40% (V_2) of the rated voltage is measured. The capacitance value is calculated as:

$$C_{\text{total}} = I_{\text{discharge}} \cdot \frac{t_2 - t_1}{V_1 - V_2}$$

The value of the discharge current is determined by the application. The IEC standard defines four classes:

- Class 1, Memory backup, discharge current in mA = 1 • C (F)
- Class 2, Energy storage, discharge current in mA = 0,4 • C (F) • V (V)
- Class 3, Power, discharge current in mA = 4 • C (F) • V (V)
- Class 4, Instantaneous power, discharge current in mA = 40 • C (F) • V (V)

The measurement methods employed by individual manufacturers are mainly comparable to the standardized methods.

The standardized measuring method is too time consuming for manufacturers to use during production for each individual component. For industrial produced capacitors the capacitance value is instead measured with a faster low frequency AC voltage and a correlation factor is used to compute the rated capacitance.

This frequency dependence affects capacitor operation. Rapid charge and discharge cycles mean that neither the rated capacitance value nor energy density are available. In this case the rated capacitance value is recalculated for each application condition.

Operating Voltage

Supercapacitors are low voltage components. Safe operation requires that the voltage remain within specified limits. The rated voltage U_R is the maximum DC voltage or peak pulse voltage that may be applied continuously and remain within the specified temperature range.

Capacitors should never be subjected to voltages continuously in excess of the rated voltage. The rated voltage includes a safety margin against the electrolyte's breakdown voltage at which the electrolyte decomposes. The breakdown voltage decomposes the separating solvent molecules in the Helmholtz double-layer, f. e. water splits into hydrogen and oxide. The solvent molecules then cannot separate the electrical charges from each other. Higher voltages than rated voltage cause hydrogen gas formation or a short circuit.

Figure: *A supercapacitor with 5.5 volts is constructed out of two single cells 2.25 volt each in series connection*

Standard supercapacitors with aqueous electrolyte normally are specified with a rated voltage of 2.1 to 2.3 V and capacitors with organic solvents with 2.5 to 2.7 V. Lithium-ion capacitors with doped electrodes may reach a rated voltage of 3.8 to 4 V, but have a lower voltage limit of about 2.2 V.

Operating supercapacitors below the rated voltage improves the long-time behaviour of the electrical parameters. Capacitance values and internal resistance during cycling are more stable and lifetime and charge/discharge cycles may be extended.

Supercapacitors rated voltages are generally lower than applications require. Higher application voltages require connecting cells in series. Since each component has a slight difference in capacitance value and ESR, it is necessary to actively or passively balance them to stabilize the applied voltage. Passive balancing employs resistors in parallel with the supercapacitors. Active balancing may include electronic voltage management above a threshold that varies the current.

Internal Resistance

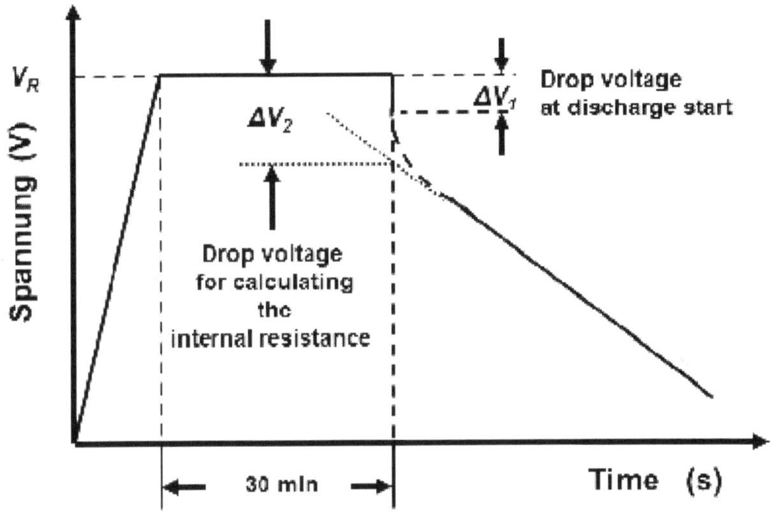

Figure: *The internal DC resistance can be calculated out of the voltage drop obtained from the intersection of the auxiliary line extended from the straight part and the time base at the time of discharge start*

Charging/discharging a supercapacitor is connected to the movement of charge carriers (ions) in the electrolyte across the separator to the electrodes and into their porous structure. Losses occur during this movement that can be measured as the internal DC resistance.

With the electrical model of cascaded, series-connected RC (resistor/capacitor) elements in the electrode pores, the internal resistance increases with the increasing penetration depth of the charge carriers into the pores. The internal DC resistance is time dependent and increases during charge/discharge. In applications often only the switch-on and switch-off range is interesting. The internal resistance R_i can be calculated from the voltage drop ΔV_2 at the time of discharge, starting with a constant discharge current $I_{discharge}$. It is obtained from the intersection of the auxiliary line extended from the straight part and the time base at the time of discharge start. Resistance can be calculated by:

$$R_i = \frac{\Delta V_2}{I_{discharge}}$$

The discharge current $I_{discharge}$ for the measurement of internal resistance can be taken from the classification according to IEC 62391-1.

This internal DC resistance R_i should not be confused with the internal AC resistance called Equivalent Series Resistance (ESR) normally specified for capacitors. It is measured at 1 kHz. ESR is much smaller than DC resistance. ESR is not relevant for calculating superconductor inrush currents or other peak currents.

R_i determines several supercapacitor properties. It limits the charge and discharge peak currents as well as charge/discharge times. R_i and the capacitance C results in the time constant *t*

$$\tau = R_i \cdot C$$

This time constant determines the charge/discharge time. A 100 F capacitor with an internal resistance of 30 mΩ for example, has a time constant of 0.03 • 100 = 3 s. After 3 seconds charging with a current limited only by internal resistance, the capacitor has 62.3% of full charge (or is discharged to 36.8% of full charge).

Standard capacitors with constant internal resistance fully charge during about 5 τ. Since internal resistance increases with charge/discharge, actual times cannot be calculated with this formula. Thus, charge/discharge time depends on specific individual construction details.

Current Load and Cycle Stability

Because supercapacitors operate without forming chemical bonds, current loads, including charge, discharge and peak currents are not limited by reaction constraints. Current load and cycle stability can be much higher than for rechargeable batteries. Current loads are limited only by internal resistance, which may be substantially lower than for batteries.

Internal resistance "R_i" and charge/discharge currents or peak currents "I" generate internal heat losses "P_{loss}" according to:

$$P_{loss} = R_i \cdot I^2$$

This heat must be released and distributed to the ambient environment to maintain operating temperatures below the specified maximum temperature.

Heat generally defines capacitor lifetime because of electrolyte diffusion. The heat generation coming from current loads should be smaller than 5 to 10 K at maximum ambient temperature (which has only minor influence on expected lifetime). For that reason the specified charge and discharge currents for frequent cycling are determined by internal resistance.

The specified cycle parameters under maximal conditions include charge and discharge current, pulse duration and frequency. They are specified for a defined temperature range and over the full voltage range for a defined lifetime. They can differ enormously depending on the combination of electrode porosity, pore size and electrolyte. Generally a lower current load increases capacitor life and increases the number of cycles. This can be achieved either by a lower voltage range or slower charging and discharging.

Supercapacitors (except those with polymer electrodes) can potentially support more than one million charge/discharge cycles without substantial capacity drops or internal resistance increases. Beneath the higher current load is this the second great advantage of supercapacitors over batteries. The stability results from the dual electrostatic and electrochemical storage principles.

The specified charge and discharge currents can be significantly exceeded by lowering the frequency or by single pulses. Heat generated by a single pulse may be spread over the time until the next pulse occurs to ensure a relatively small average heat increase. Such a "peak power current" for power applications for supercapacitors of more than 1000 F can provide a maximum peak current of about 1000 A. Such high currents generate high thermal stress and high electromagnetic forces that can damage the electrode-collector connection requiring robust design and construction of the capacitors.

Energy Density and Power Density

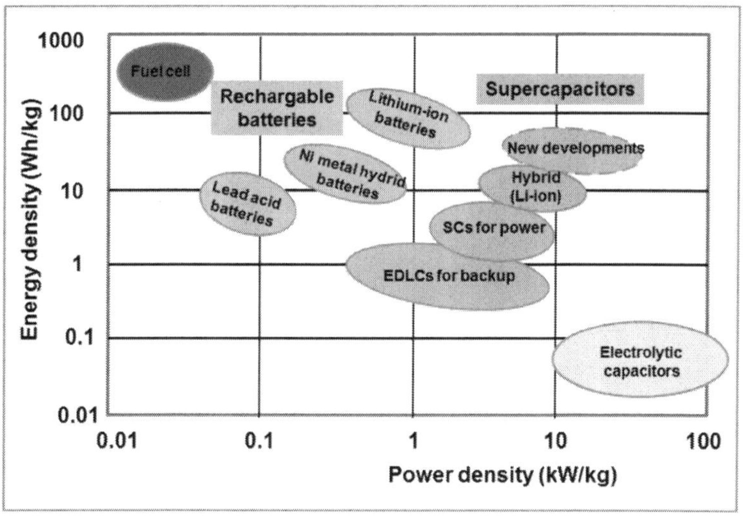

Figure: *Ragone chart showing power density vs. energy density of various capacitors and batteries*

Supercapacitors occupy the gap between high power/low energy electrolytic capacitors and low power/high energy rechargeable batteries.

The amount of energy stored in a supercapacitor is called specific energy. The energy W_{max} of a capacitor is given by the formula

$$W_{max} = \frac{1}{2} \cdot C_{total} \cdot V_{loaded}^2$$

This formula describes the amount of energy stored and is often used to describe new research successes. However, only part of the stored energy is available to applications, because the voltage drop and the time constant over the internal resistance mean that some of the stored charge is inaccessible. The effective realised amount of energy W_{eff} is reduced by the used voltage difference between V_{max} and V_{min} and can be represented as:

$$W_{eff} = \frac{1}{2} C \cdot (V_{max}^2 - V_{min}^2)$$

This formula also represents the energy asymmetric voltage components such as lithium ion capacitors.

Energy density is either measured gravimetrically (per unit of mass) in watt-hours per kilogram (Wh/kg) or volumetrically (per unit of volume) in watt-hours per litre (Wh/l).

As of 2013 commercial gravimetric energy densities range from around 0.5 to 15 Wh/kg. For comparison, an aluminum electrolytic capacitor stores typically 0.01 to 0.3 Wh/kg, while a conventional lead-acid battery stores typically 30 to 40 Wh/kg and modern lithium-ion batteries 100 to 265 Wh/kg. Supercapacitors can therefore store 10 to 100 times more energy than electrolytic capacitors, but only one tenth as much as batteries.

Although the energy densities of supercapacitors are insufficient compared with batteries the capacitors have an important advantage, the power density. Power density describes the speed at which energy can be delivered to/absorbed from the load. The maximum power P_{max} is given by the formula:

$$P_{max} = \frac{1}{4} \cdot \frac{V^2}{R_i}$$

with V = voltage applied and R_i, the internal DC resistance.

Power density is measured either gravimetrically in kilowatts per kilogram (kW/kg) or volumetrically in kilowatts per litre (kW/l).

The described maximum power P_{max} specifies the power of a theoretical rectangular single maximum current peak of a given voltage. In real circuits the current peak is not rectangular and the voltage is smaller, caused by the voltage drop. IEC 62391–2 established a more realistic effective power P_{eff} for supercapacitors for power applications:

$$P_{\text{eff}} = \frac{1}{8} \cdot \frac{V^2}{R_i}$$

Supercapacitor power density is typically 10 to 100 times greater than for batteries and can reach values up to 15 kW/kg.

Ragone charts relate energy to power and are a valuable tool for characterizing and visualizing energy storage components. With such a diagram, the position of power density and energy density of different storage technologies is easily to compare.

Lifetime

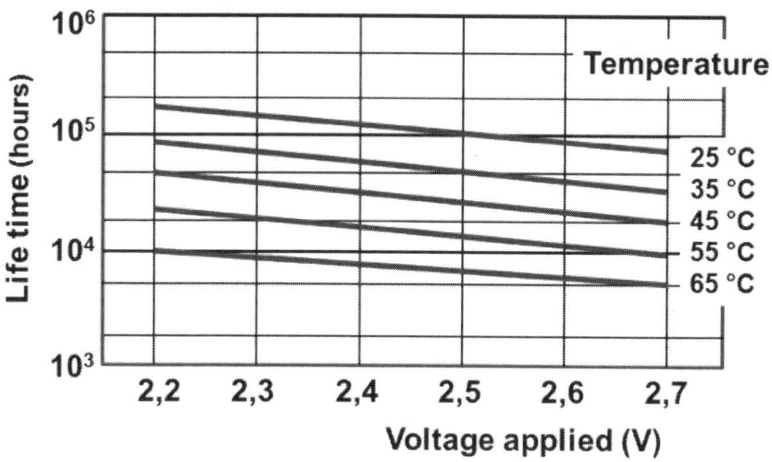

Figure: *The lifetime of supercapacitors depends mainly on the capacitor temperature and the voltage applied*

Supercapacitors exhibit a much longer lifetime than batteries. Since supercapacitors do not rely on chemical changes in the electrodes (except for those with polymer electrodes) lifetimes depend mostly on the rate of evaporation of the liquid electrolyte. This evaporation in general is a function of temperature, of current load, current cycle frequency and voltage. Current load and cycle frequency generate internal heat, so that the evaporation-determining temperature is the sum of ambient and internal heat. This temperature is measurable as core temperature in the centre of a capacitor body. The higher the core temperature the faster the evaporation and the shorter the lifetime.

Evaporation generally results in decreasing capacitance and increasing internal resistance. According to IEC/EN 62391-2 capacitance reductions of over 30% or internal resistance exceeding four times its data sheet specifications are considered “wear-out failures”, implying that the component has reached end-of-life. The capacitors are operable, but with reduced capabilities. It depends on the application of the capacitors, whether the aberration of the parameters have any influence on the proper functionality or not.

Such large changes of electrical parameters specified in IEC/EN 62391-2 are usually unacceptable for high current load applications. Components that support high current loads use much smaller limits, e.g., 20% loss of capacitance or double the internal resistance. The narrower definition is important for such applications, since heat increases linearly with increasing internal resistance and the maximum temperature should not be exceeded. Temperatures higher than specified can destroy the capacitor.

The real application lifetime of supercapacitors, also called “service life”, “life expectancy” or “load life”, can reach 10 to 15 years or more at room temperature. Such long periods cannot be tested by manufacturers. Hence, they specify the expected capacitor lifetime at the maximum temperature and voltage conditions. The results are specified in datasheets using the notation “tested time (hours)/max. temperature (°C)”, such as “5000 h/65 °C”. With this value and a formula, lifetimes can be estimated for lower conditions.

Datasheet lifetime specification is tested by the manufactures using an accelerated aging test called “endurance test” with maximum temperature and voltage over a specified time. For a “zero defect” product policy during this test no wear out or total failure may occur.

The lifetime specification from datasheets can be used for estimation of expected lifetime according to conditions coming from the application. The “10-degrees-rule” used for electrolytic capacitors with non-solid electrolyte is used for those estimations and can be used for supercapacitors, too. This rule employs the Arrhenius equation, a simple formula for the temperature dependence of reaction rates. For every 10 °C reduction in operating temperature, the estimated life doubles.

$$L_x = L_0 \cdot 2^{\frac{T_0 - T_x}{10}}$$

With

- L_x = estimated lifetime
- L_0 = specified lifetime
- T_0 = upper specified capacitor temperature
- T_x = actual operating temperature of the capacitor cell

Calculated with this formula, capacitors specified with 5000 h at 65 °C, have an estimated lifetime of 20,000 h at 45 °C.

Lifetimes are also dependent on the operating voltage, because the development of gas in the liquid electrolyte depends on the voltage. The lower the voltage the smaller the gas development and the longer the lifetime. No general formula relates voltage to lifetime. The voltage dependent curves shown from the picture are an empirical result from one manufacturer.

Life expectancy for power applications may be also limited by current load or number of cycles. This limitation has to be specified by the relevant manufacturer and is strongly type dependent.

Self-Discharge

Storing electrical energy in the double-layer separates the charge carriers by distance within the pores by distances in the range of molecules. Over this short distance irregularities can occur, leading to a small exchange of charge carriers and gradual discharge. This self-discharge is called leakage current. Leakage depends on capacitance, voltage, temperature and the chemical stability of the electrode/electrolyte combination. At room temperature leakage is so low that it is specified as time to self-discharge. Supercapacitor self-discharge time is specified in hours, days or weeks. As an example, a 5.5 V/1 F Panasonic "Goldcapacitor" specifies a voltage drop at 20 °C from 5.5 V down to 3 V in 600 hours (25 days or 3.6 weeks) for a double cell capacitor.

Polarity

Figure: *A negative bar on the insulating sleeve indicates the cathode terminal of the capacitor*

Although the anode and cathode of symmetric supercapacitors consist of the same material, theoretically supercapacitors have no true polarity. Normally catastrophic failure does not occur, however reverse-charging a supercapacitor lowers its capacity. It is recommended practice to maintain the polarity resulting from a formation of the electrodes during production. Asymmetric supercapacitors are inherently polar.

Supercapacitors may not be operated with reverse polarity, precluding AC operation.

A bar in the insulating sleeve identifies the cathode terminal in a polarized component.

The terms "anode" and "cathode" can lead to confusion, because the polarity changes depending on whether a component is considered as a generator or as a consumer. For an accumulator or a battery the cathode has a positive polarity (+) and the anode has negative polarity (-). For capacitors the cathode has negative polarity (-) and the anode has positive polarity (+). This requires special attention if supercapacitors are substituted or switched in parallel with batteries.

Comparison of Technical Parameters

Component Comparison

Mixing electrodes and electrolytes yields a variety of components suitable for diverse applications. The development of low-ohmic electrolyte systems, in combination with electrodes with high pseudocapacitance, enable many more technical solutions.

The following table shows differences among capacitors of various manufacturers in capacitance range, cell voltage, internal resistance (ESR, DC or AC value) and volumetric and gravimetric energy density.

In the table, ESR refers to the component with the largest capacitance value of the respective manufacturer. Roughly, they divide supercapacitors into two groups. The first group offers greater ESR values of about 20 milliohms and relatively small capacitance of 0.1 to 470 F. These are "double-layer capacitors" for memory back-up or similar applications. The second group offers 100 to 10,000 F with a significantly lower ESR value under 1 milliohm. These components are suitable for power applications. A correlation of some supercapacitor series of different manufacturers to the various construction features is provided in Pandolfo and Hollenkamp.

Electrolytic capacitors feature unlimited charge/discharge cycles, high dielectric strength (up to 550 V) and good frequency response as AC resistance in the lower frequency range. Supercapacitors can store

10 to 100 times more energy than electrolytic capacitors but they do not support AC applications.

With regards to rechargeable batteries supercapacitors feature higher peak currents, low cost per cycle, no danger of overcharging, good reversibility, non-corrosive electrolyte and low material toxicity, while batteries offer, lower purchase cost, stable voltage under discharge, but they require complex electronic control and switching equipment, with consequent energy loss and spark hazard given a short.

Standards

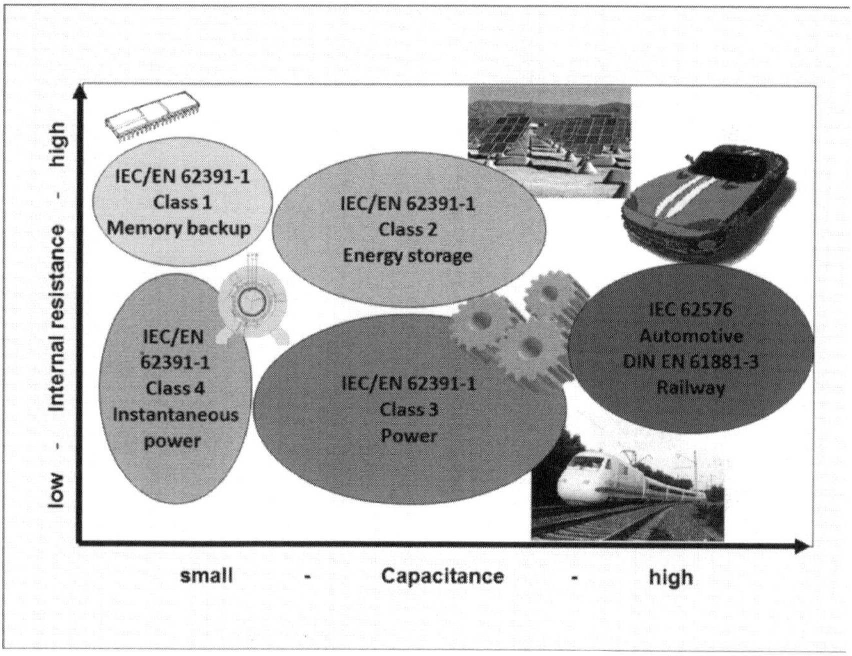

Figure: *Classification of supercapacitors into classes regarding to IEC 62391-1, IEC 62567 and BS EN 61881-3 standards*

Supercapacitors vary sufficiently that they are rarely interchangeable, especially those with higher energy densities. Applications range from low to high peak currents, requiring standardized test protocols.

Test specifications and parameter requirements are specified in the generic specification

- IEC/EN 62391–1, *Fixed electric double layer capacitors for use in electronic equipment.*

The standard defines four application classes, according to discharge current levels:

- Class 1: Memory backup
- Class 2: Energy storage, mainly used for driving motors require a short time operation,
- Class 3: Power, higher power demand for a long time operation,
- Class 4: Instantaneous power, for applications that requires relatively high current units or peak currents ranging up to several hundreds of amperes even with a short operating time

Three further standards describe special applications:

- IEC 62391–2, *Fixed electric double-layer capacitors for use in electronic equipment - Blank detail specification - Electric double-layer capacitors for power application -*
- IEC 62576, *Electric double-layer capacitors for use in hybrid electric vehicles. Test methods for electrical characteristics*
- BS/EN 61881-3, *Railway applications. Rolling stock equipment. Capacitors for power electronics. Electric double-layer capacitors*

Applications

Supercapacitors do not support AC applications.

Supercapacitors have advantages in applications where a large amount of power is needed for a relatively short time, where a very high number of charge/discharge cycles or a longer lifetime is required. Typical applications range from milliamp currents or milliwatts of power for up to a few minutes to several amps current or several hundred kilowatts power for much shorter periods.

The time t a supercapacitor can deliver a constant current I can be calculated as:

$$t = \frac{C \cdot (U_{\text{charge}} - U_{\min})}{I}$$

as the capacitor voltage decreases from U_{charge} down to U_{min}.

If the application needs a constant power P for a certain time t this can be calculated as:

$$t = \frac{1}{2P} \cdot C \cdot (U_{\text{charge}}^2 - U_{\min}^2).$$

wherein also the capacitor voltage decreases from U_{charge} down to U_{min}.

General Applications

Consumer Applications: In applications with fluctuating loads, such as laptop computers, PDA's, GPS, portable media players, hand-held devices, and photovoltaic systems, supercapacitors can stabilize the power supply.

A cordless electric screwdriver with supercapacitors for energy storage has about half the run time of a comparable battery model, but can be fully charged in 90 seconds. It retains 85% of its charge after three months left idle. Supercapacitors deliver power for photographic flashes in digital cameras and for LED life flashlights that can be charged in, e.g., 90 seconds. As of 2013, portable speakers powered by supercapacitors were offered to the market.

Industrial Applications

Supercapacitors provide backup or emergency shutdown power to low-power equipment such as RAM, SRAM, micro-controllers and PC Cards. They are the sole power source for low energy applications such as automated metre reading (AMR) equipment or for event notification in industrial electronics. Supercapacitors buffer power to and from rechargeable batteries, mitigating the effects of short power interruptions and high current peaks. Batteries kick in only during extended interruptions, e.g., if the mains power or a fuel cell fails, which lengthens battery life. Typical industrial applications for such circuits are uninterruptible power supplies (UPS), where supercapacitors have replaced much larger banks of electrolytic capacitors. This combination reduces the cost per cycle, saves on replacement and maintenance costs, enables the battery to be downsized and extends battery life. A disadvantage is the need for a special circuit to reconcile the differing behaviours.

Supercapacitors are used as battery replacement in some pit trains in China to substitute conventional trolleys in coal mines. They bring coal to the surface. This approach removes a fire and safety hazard from coal mines coming from the batteries. The capacitors can be charged at the surface in less than 30 minutes.

Renewable Energy

Figure: *Rotor with wind turbine pitch system*

Supercapacitors provide backup power for actuators in wind turbine pitch systems, so that blade pitch can be adjusted even if the main supply fails.

Supercapacitors can stabilize voltage for powerlines. Wind and photovoltaic systems exhibit fluctuating loads evoked by clouds that supercapacitors can buffer within milliseconds. This helps stabilize grid voltage and frequency, balance supply and demand of power and manage real or reactive power.

Energy Harvesting

Supercapacitors are suitable temporary energy storage devices for energy harvesting systems. In energy harvesting systems the energy is collected from the ambient or renevable sources, e.g. mechanical movement, light or electromagnetic fields, and converted to electrical energy in an energy storage device. For example, it was demonstrated that energy collected from RF (radio frequency) field using an RF antenna as appropriate rectifier circuit can be stored to a printed supercapacitor, and the harvested energy was then used to power an application-specific integrated circuit (ASIC) circuit for over 10 hours.

Public Sector

Figure: *Street light combining a solar cell power source with LED lamps and supercapacitors for energy storage*

Sado City, in Japan's Niigata Prefecture, has street lights that combine a stand-alone power source with solar cells and LEDs. Supercapacitors store the solar energy and supply 2 LED lamps, providing 15 W power consumption overnight. The supercapacitors can last more than 10 years and offer stable performance under various weather conditions, including temperatures from +40 to below -20 °C.

Aerial Lift

***Figure:** Aerial lift in Zell am See, Austria*

In Zell am See, Austria, an aerial lift connects the city with Schmittenhöhe mountain. The gondolas sometimes run 24 hours per day, using electricity for lights, door opening and communication. The only available time for recharging batteries at the stations is during the brief intervals of guest loading and unloading, which is too short to recharge batteries. Supercapacitors offer a fast charge, higher number of cycles and longer life time than batteries.

Emirates Air Line (cable car), also known as the Thames cable car, is a 1-kilometre (0.62 mi) gondola line that crosses the Thames from the Greenwich Peninsula to the Royal Docks. The cabins are equipped with a modern infotainment system, which is powered by supercapacitors.

Medical

Supercapacitors are used in defibrillators where they can deliver 500 joules to shock the heart back into sinus rhythm.

Aviation

In 2005, aerospace systems and controls company Diehl Luftfahrt Elektronik GmbH chose supercapacitors to power emergency actuators for doors and evacuation slides used in airliners, including the Airbus 380.

Military

Supercapacitors' low internal resistance supports applications that require short-term high currents. Among the earliest uses were motor startup (cold diesel engine start) for large engines in tanks and submarines. Supercapacitors buffer the battery, handling short current peaks and reducing cycling. Further military applications that require high power density are phased array radar antennae, laser power supplies, military radio communications, avionics displays and instrumentation, backup power for airbag deployment and GPS-guided missiles and projectiles.

Energy Recovery

A primary challenge of all transport is reducing energy consumption and reducing CO_2 emissions. Recovery of braking energy (recuperation or regeneration) helps with both. This requires components that can quickly store and release energy over long times with a high cycle rate. Supercapacitors fulfill these requirements and are therefore used in a lot of applications in all kinds of transportation.

Railway

Figure: *Green Cargo operates TRAXX locomotives from Bombardier Transportation*

Supercapacitors can be used to supplement batteries in starter systems in diesel railroad locomotives with diesel-electric transmission. The capacitors capture the braking energy of a full stop and deliver the peak current for starting the diesel engine and acceleration of the train and ensures the stabilization of catenary voltage. Depending on the driving mode up to 30% energy saving is possible by recovery of breaking energy. Low maintenance and environmentally friendly materials encouraged the choice of supercapacitors.

Cranes, Forklifts and Tractors

Mobile hybrid diesel-electric rubber tyred gantry cranes move and stack containers within a terminal. Lifting the boxes requires large amounts of energy. Some of the energy could be recaptured while lowering the load resulting in improved efficiency.

Figure: *Container Yard with Rubber Tyre Gantry Crane*

A triple hybrid forklift truck uses fuel cells and batteries as primary energy storage and supercapacitors to buffer power peaks by storing braking energy. They provide the fork lift with peak power over 30 kW. The triple-hybrid system offers over 50% energy savings compared with diesel or fuel-cell systems.

Supercapacitor-powered terminal tractors transport containers to warehouses. They provide an economical, quiet and pollution-free alternative to diesel terminal tractors.

Light-Rails and Trams

Supercapacitors make it possible not only to reduce energy but additional to do away catenary overhead lines in historical city areas preserving the city's architectural heritage. This approach may allow many new LRV city lines to serve catenary overhead wires that are too expensive to fully route installation. In 2003 Mannheim, adopted a prototype light-rail vehicle (LRV) using the MITRAC Energy Saver system from Bombardier Transportation to store mechanical braking energy with a roof-mounted supercapacitor unit. It contains several units each made of 192 capacitors with 2700 F /2.7 V interconnected in three parallel lines.

Figure: *Light rail vehicle in Mannheim*

This circuit results in a 518 V system with an energy content of 1.5 kWh. For acceleration when starting this "on-board-system" can provided the LRV with 600 kW and can drive the vehicle up to 1 km without catenary supply integrating the LRV into the urban environment by driving without catenary lines. Compared to conventional LRVs or Metro vehicles that return energy into the grid, onboard energy storage saves up to 30% and reduces peak grid demand by up to 50%.

Figure: *The Paris T3 line runs without catenary overhead wires in some sections*

In 2009 supercapacitors enabled LRV's to operate in the historical city area of Heidelberg without catenary overhead wires preserving the city's architectural heritage.

The SC equipment cost an additional €270,000 per vehicle, which was expected to be recovered over the first 15 years of operation. The supercapacitors are charged at stop-over stations when the vehicle is at a scheduled stop. This approach may allow many LRV city lines to serve catenary overhead wires that are too expensive to fully route installation. In April 2011 German regional transport operator Rhein-Neckar, responsible for Heidelberg, ordered a further 11 units.

In 2009 in Paris a tram on route T3 operates with an energy recovery system of the manufacturer Alstom called "STEEM". The system is fitted with 48 roof-mounted supercapacitors to store braking energy provides tramways with a high level of energy autonomy by enabling them to run without catenary power on parts of its route, recharging while travelling on powered stop-over stations. During the tests, the tramset used an average of approximately 16% less energy.

Figure: *Light rail vehicle in Hong Kong*

In 2012 tram operator Geneva Public Transport began tests of an LRV equipped with a prototype roof-mounted supercapacitor unit to recover braking energy.

Siemens is delivering supercapacitor-enhanced light-rail transport systems that include mobile storage.

Hong Kong's South Island metro line is to be equipped with two 2 MW energy storage units that are expected to reduce energy consumption by 10%.

In August 2012 the CSR Zhouzhou Electric Locomotive corporation of China presented a prototype two-car light metro train equipped with a roof-mounted supercapacitor unit. The train can travel up 2 km without wires, recharging in 30 seconds at stations via a ground mounted pickup. The supplier claimed the trains could be used in 100 small and medium-sized Chinese cities.

In 2012, in Lyon (France), the SYTRAL (Lyon public transportation administration) started experiments of a "way side regeneration" system built by Adetel Group which has developed its own energy saver named 3 NeoGreen3 for LRV, LRT and metros.

Buses

Figure: *MAN Ultracapbus in Nuremberg, Germany*

The first hybrid bus with supercapacitors in Europe came in 2001 in Nuremberg, Germany. It was MAN's so-called "Ultracapbus", and was tested in real operation in 2001/2002. The test vehicle was equipped with a diesel-electric drive in combination with supercapacitors. The system was supplied with 8 Ultracap modules of 80 V, each containing 36 components. The system worked with 640 V and could be charged/ discharged at 400 A. Its energy content was 0.4 kWh with a weight of 400 kg.

The supercapacitors recaptured braking energy and delivered starting energy. Fuel consumption was reduced by 10 to 15% compared to conventional diesel vehicles. Other advantages included reduction of CO_2 emissions, quiet and emissions-free engine starts, lower vibration and reduced maintenance costs.

Figure: *Electric bus at EXPO 2010 in Shanghai (Capabus) recharging at the bus stop*

As of 2002 in Luzern, Switzerland an electric bus fleet called TOHYCO-Rider was tested. The supercapacitors could be recharged via an inductive contactless high-speed power charger after every transportation cycle, within 3 to 4 minutes.

In early 2005 Shanghai tested a new form of electric bus called capabus that runs without powerlines (catenary free operation) using large onboard supercapacitors that partially recharge whenever the bus is at a stop (under so-called electric umbrellas), and fully charge in the terminus. In 2006, two commercial bus routes began to use the capabuses; one of them is route 11 in Shanghai. It was estimated that the supercapacitor bus was cheaper than a lithium-ion battery bus, and one of its buses had one-tenth the energy cost of a diesel bus with lifetime fuel savings of $200,000.

A hybrid electric bus called tribrid was unveiled in 2008 by the University of Glamorgan, Wales, for use as student transport. It is powered by hydrogen fuel or solar cells, batteries and ultracapacitors.

Motor Racing

The FIA, a governing body for motor racing events, proposed in the *Power-Train Regulation Framework for Formula 1* version 1.3 of 23 May 2007 that a new set of power train regulations be issued that includes a hybrid drive of up to 200 kW input and output power using "superbatteries" made with batteries and supercapacitors connected in parallel (KERS). About 20% tank-to-wheel efficiency could be reached using the KERS system.

Figure: *World champion Sebastian Vettel in Malaysia 2010*

Figure: *Toyota TS030 Hybrid at 2012 24 Hours of Le Mans motor race*

The Toyota TS030 Hybrid LMP1 car, a racing car developed under Le Mans Prototype rules, uses a hybrid drivetrain with supercapacitors. In the 2012 24 Hours of Le Mans race a TS030 qualified with a fastest lap only 1.055 seconds slower (3:24.842 versus 3:23.787) than the fastest car, an Audi R18 e-tron quattro with flywheel energy storage. The supercapacitor and flywheel components, whose rapid charge-discharge capabilities help in both braking and acceleration, made the Audi and Toyota hybrids the fastest cars in the race. In the 2012 Le Mans race the two competing TS030s, one of which was in the lead for part of the race, both retired for reasons unrelated to the supercapacitors. The TS030 won three of the 8 races in the 2012 FIA World Endurance Championship season.

Hybrid Electric Vehicles

Figure: *Mazda2 (since 2010)*

Supercapacitor/battery combinations in electric vehicles (EV) and hybrid electric vehicles (HEV) are well investigated. A 20 to 60% fuel reduction has been claimed by recovering brake energy in EVs or HEVs. The ability of supercapacitors to charge much faster than batteries, their stable electrical properties, broader temperature range and longer lifetime are suitable, but weight, volume and especially cost mitigate those advantages.

Supercapacitors lower energy density makes them unsuitable for use as a stand-alone energy source for long distance driving. The fuel economy improvement between a capacitor and a battery solution is about 20% and is available only for shorter trips. For long distance driving the advantage decreases to 6%. Vehicles combining capacitors and batteries run only in experimental vehicles.

As of 2013 all automotive manufacturers of EV or HEVs have developed prototypes that uses supercapacitors instead of batteries to store braking energy in order to improve driveline efficiency. Only the Mazda 6 uses supercapacitors to recover braking energy called i-eloop to reduce fuel consumption about 10%.

Russian Yo-cars ¨-mobile series is an ¸-concept and ¸-crossover hybrid vehicle working with a gas driven Wankel motor and an electric generator for driving. A supercapacitor with relatively low capacitance recovers brake energy to power the electric motor when accelerating from a stop.

New Developments

As of 2013 commercially available lithium-ion supercapacitors offered the highest gravimetric energy density to date, reaching 15 Wh/kg. Research focuses on improving energy density, reducing internal resistance, expanding temperature range, increasing lifetimes and reducing costs. Projects include nanostructured electrodes, tailored-pore-size electrodes, pseudocapacitive coating or doping materials and improved electrolytes.

As of 2010 worldwide sales of supercapacitors reached US$400 million.

The market for batteries (estimated by Frost & Sullivan) grew from US$47.5 billion, (76.4% or US$36.3 billion of which was rechargeable batteries) to US$95 billion. The market for supercapacitors is still a small niche market that is not keeping pace with its larger rival.

In 2012, NanoMarkets forecast sales to grow to US$3.5 billion by 2020, an increase of about 900% within 10 years. Assumptions underlying this growth include a rapidly improving price/performance ratio and evolving "green energy" applications, such as energy recovery in electric vehicles. Otherwise the market was forecast to grow about 30% overall between 2013 and 2018 and remain in the hundreds of millions of dollars.

Supercapacitor costs in 2006 were US$0.01 per farad or US$2.85 per kilojoule, moving in 2008 below US$0.01 per farad, and were expected to drop further in the medium term.

7

Environmental Applications of Nanomaterials

The environmental impact of nanotechnology is the possible effects that the use of nanotechnological materials and devices will have on the environment. As nanotechnology is an emerging field, there is great debate regarding to what extent industrial and commercial use of nanomaterials will affect organisms and ecosystems.

Nanotechnology's environmental impact can be split into two aspects: the potential for nanotechnological innovations to help improve the environment, and the possibly novel type of pollution that nanotechnological materials might cause if released into the environment.

Buckminsterfullerene

Buckminsterfullerene (or bucky-ball) is a spherical fullerene molecule with the formula C_{60}. It has a cage-like fused-ring structure (truncated icosahedron) which resembles a soccer ball, made of twenty hexagons and twelve pentagons, with a carbon atom at each vertex of each polygon and a bond along each polygon edge.

It was first generated in 1985 by Harold Kroto, James R. Heath, Sean O'Brien, Robert Curl, and Richard Smalley at Rice University. Kroto, Curl and Smalley were awarded the 1996 Nobel Prize in Chemistry for their roles in the discovery of buckminsterfullerene and the related class of molecules, the fullerenes. The name is a reference to Buckminster

Fuller, as C_{60} resembles his trademark geodesic domes. Buckminsterfullerene is the most commonly naturally occurring fullerene molecule, as it can be found in small quantities in soot. Solid and gaseous forms of the molecule have been detected in deep space.

Buckminsterfullerene is one of the largest objects to have been shown to exhibit wave–particle duality. Its discovery led to the exploration of a new field of chemistry, involving the study of fullerenes.

Buckminsterfullerene derives from the name of the noted futurist and inventor Buckminster Fuller. One of his designs of a geodesic dome structure bears great resemblance to C_{60}; as a result, the discoverers of the allotrope named the newfound molecule after him. The general public, however, sometimes refers to buckminsterfullerene, and even Mr. Fuller's dome structure, as buckyballs.

History

The structure associated with fullerenes was described by Leonardo da Vinci. Albrecht Dürer also reproduced a similar icosahedron containing 12 pentagonal and 20 hexagonal faces but there are no clear documentations of this.

Discovery

Theoretical predictions of buckyball molecules appeared in the late 1960s – early 1970s, but they went largely unnoticed. In the early 1970s, the chemistry of unsaturated carbon configurations was studied by a group at the University of Sussex, led by Harry Kroto and David Walton. In the 1980s a technique was developed by Richard Smalley and Bob Curl at Rice University, Texas to isolate these substances. They used laser vaporization of a suitable target to produce clusters of atoms. Kroto realised that by using a graphite target, any carbon chains formed could be studied. Another interesting fact is that, at the same time, astrophysicists were working along with spectroscopists to study infrared emissions from giant red carbon stars. Smalley and team were able to use a laser vaporization technique to create carbon clusters which could potentially emit infrared at the same wavelength as had been emitted by the red carbon star. Hence, the inspiration came to Smalley and team to use the laser technique on graphite to create the first fullerene molecule.

C_{60} was discovered in 1985 by Robert Curl, Harold Kroto and Richard Smalley. Using laser evaporation of graphite they found C_n clusters (where $n>20$ and even) of which the most common were C_{60} and C_{70}. The figure below shows the model of the pulsed supersonic nozzle used to create the carbon cluster beams. A solid rotating graphite disk was used as the surface from which carbon was vaporized using a laser beam creating hot plasma that was then passed through a stream of high-density helium gas. The carbon species were subsequently cooled and ionized resulting in the formation of clusters. Clusters ranged in molecular masses but Kroto and Smalley found predominance in a C_{60} cluster that could be enhanced further by letting the plasma react longer.[3, 6] They also discovered that the C_{60} molecule formed a cage-like structure, a regular truncated icosahedron.

For this discovery they were awarded the 1996 Nobel Prize in Chemistry. The discovery of buckyballs was surprising, as the scientists aimed the experiment at producing carbon plasmas to replicate and characterize unidentified interstellar matter. Mass spectrometry analysis of the product indicated the formation of spheroidal carbon molecules.

The experimental evidence, a strong peak at 720 atomic mass units, indicated that a carbon molecule with 60 carbon atoms was forming, but provided no structural information. The research group concluded after reactivity experiments, that the most likely structure was a spheroidal molecule. The idea was quickly rationalized as the basis of an icosahedral symmetry closed cage structure. Kroto mentioned geodesic dome structures of the noted futurist and inventor Buckminster Fuller as influences in the naming of this particular substance as buckminsterfullerene.

Further Developments

The versatility of fullerene molecules has led to a large amount of research exploring their properties. One interesting property is fullerene's large-capacity internal spaces. Atoms of different elements may be placed inside the molecular cage formed by the carbon atoms, producing a shrink-wrapped version of these elements.

Beam-experiments conducted between 1985 and 1990 provided more evidence for the stability of C_{60} while supporting the closed-cage structural theory and predicting some of the bulk properties such a molecule would have. Around this time, intense theoretical group theory activity also predicted that C_{60} should have only four IR-active vibrational bands, on account of its icosahedral symmetry.

In 1989 physicists Wolfgang Krätschmer and Donald R. Huffman observed unusual optical absorptions in thin carbon films produced by arc-processed graphite rods. Among other features, the IR spectra showed four discrete bands in close agreement to those proposed for C_{60}. A paper published by the group in 1990 followed on from their thin film experiments, and detailed the extraction of a benzene soluble material from the arc-processed graphite. This extract had crystal and X-ray analysis consistent with arrays of spherical C_{60} molecules, approximately 0.7 nm in diameter.

Synthesis

Figure: *High-vacuum electrolysis of a C_{60}-fullerene derivative. Slow diffusion into the anode (right side) yields the characteristic purple colour of pure C_{60}.*

In 1990, W. Krätschmer and D. R. Huffman developed a simple and efficient method of producing fullerenes in gram and even kilogram amounts, which has boosted the fullerene research. In this technique, carbon soot is produced from two high-purity graphite electrodes by igniting an arc discharge between them in an inert atmosphere (helium gas). Alternatively, soot is produced by laser ablation of graphite or pyrolysis of aromatic hydrocarbons. Fullerenes are extracted from the soot using a multistep procedure. First, the soot is dissolved in appropriate organic solvents. This step yields a solution containing up to 75% of C_{60}, as well as other fullerenes. These fractions are separated using chromatography. Generally, the fullerenes are dissolved in hydrocarbon or halogenated hydrocarbon and separated using alumina columns.

Properties

Molecule: The structure of a buckminsterfullerene is a truncated icosahedron with 60 vertices and 32 faces (20 hexagons and 12 pentagons where no pentagons share a vertex) with a carbon atom at the vertices of each polygon and a bond along each polygon edge. The van der Waals diameter of a C_{60} molecule is about 1.01 nanometres (nm). The nucleus to nucleus diameter of a C_{60} molecule is about 0.71 nm. The C_{60} molecule has two bond lengths. The 6:6 ring bonds (between two hexagons) can be considered "double bonds" and are shorter than the 6:5 bonds (between a hexagon and a pentagon). Its average bond length is 0.14 nm. Each carbon atom in the structure is bonded covalently with 3 others.

The C_{60} molecule is extremely stable, withstanding high temperatures and high pressures. The exposed surface of the structure can selectively react with other species while maintaining the spherical geometry. Atoms and small molecules can be entrapped at the interior without reacting.

C_{60} undergoes six reversible, one-electron reductions to C6–60, whereas oxidation is irreversible. The first reduction requires is ~1.0 V (Fc/Fc+), indicating that C_{60} is a moderately effective electron acceptor. C60 has a tendency of avoiding having double bonds within the pentagonal rings which makes electron delocalization poor, and results in the fact that C_{60} is not "superaromatic". C_{60} behaves very much like an electron deficient alkene and readily reacts with electron rich species.

A carbon atom in the C60 molecule can be substituted by a nitrogen or boron atom yielding a C59N or $C_{59}B$ respectively.

Solution

Saturated Solubility of C_{60}(S, mg/mL)Solvent	***S***
1-chloronaphthalene	51
1-methylnaphthalene	33
1,2-dichlorobenzene	24
1,2,4-trimethylbenzene	18
tetrahydronaphthalene	16
carbon disulfide	8
1,2,3-tribromopropane	8
xylene	5
bromoform	5
cumene	4
toluene	3
benzene	1.5
carbon tetrachloride	0.447
chloroform	0.25
n-hexane	0.046

Contd...

Saturated Solubility of C_{60}(S, mg/mL)Solvent	***S***
cyclohexane	0.035
tetrahydrofuran	0.006
acetonitrile	0.004
methanol	0.00004
water	1.3×10^{-11}
pentane	0.004
octane	0.025
isooctane	0.026
decane	0.070
dodecane	0.091
tetradecane	0.126
dioxane	0.0041
mesitylene	0.997
dichloromethane	0.254

Fullerenes are sparingly soluble in aromatic solvents such as toluene and carbon disulfide. It is insoluble in water. Solutions of pure C_{60} have a deep purple colour which leaves a brown residue upon evaporation. The reason for this colour change is the relatively narrow energy width of the band of molecular levels responsible for green light absorption by individual C_{60} molecules. Thus individual molecules transmit some blue and red light resulting in a purple colour. Upon drying, intermolecular interaction results in the overlap and broadening of the energy bands, thereby eliminating the blue light transmittance and causing the purple to brown colour change.

C_{60} crystallises with some solvents in the lattice ("solvates"). For example, crystallization of C_{60} in benzene solution yields triclinic crystals with the formula $C_{60}\cdot 4C_6H_6$. Like other solvates, this one readily releases benzene to give the usual fcc C_{60}. Millimetre-sized crystals of C_{60} and C_{70} can be grown from solution both for solvates and for pure fullerenes.

Solid

In solid buckminsterfullerene, the molecules C_{60} stick together via the van der Waals forces in the fcc motif. At low temperatures the individual molecules are locked against rotation. Upon heating, they start rotating at about "20 °C that results in a first-order phase transition to a face-centred cubic (fcc) structure and a small, yet abrupt increase in the lattice constant from 1.411 to 1.4154 nm.

Figure: *C_{60} solid*

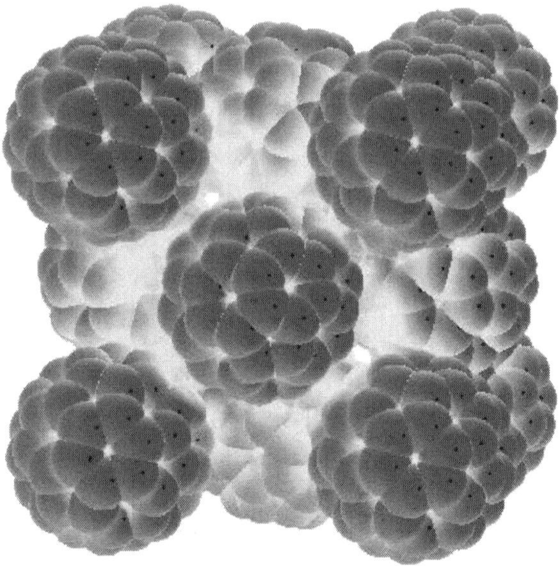

Figure: *C60 crystal structure*

C60 solid is as soft as graphite, but when compressed to less than 70% of its volume it transforms into a superhard form of diamond. C60 films and solution have strong non-linear optical properties; in particular, their optical absorption increases with light intensity (saturable absorption).

C60 forms a brownish solid with an optical absorption threshold at ~1.6 eV. It is an n-type semiconductor with a low activation energy of 0.1–0.3 eV; this conductivity is attributed to intrinsic or oxygen-related defects. Fcc C_{60} contains voids at its octahedral and tetrahedral sites which are sufficient large (0.6 and 0.2 nm respectively) to accommodate impurity atoms. When alkali metals are doped into these voids, C_{60} converts from a semiconductor into a conductor or even superconductor.

Band Structure and Superconductivity

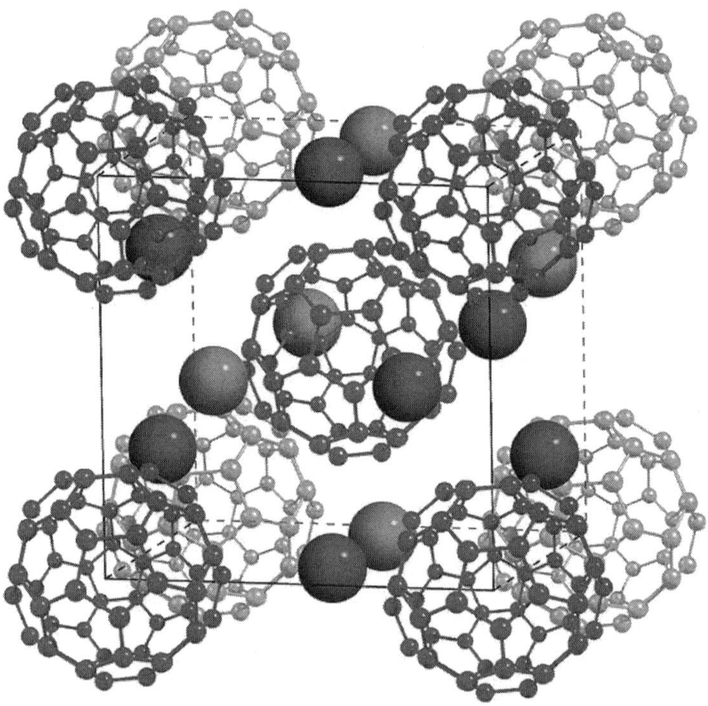

Figure: *Cs_3C_{60} crystal structure*

In 1991, Haddon *et al.* found that intercalation of alkali-metal atoms in solid C_{60} leads to metallic behaviour. In 1991, it was revealed that potassium-doped C_{60} becomes superconducting at 18 K. This was the highest transition temperature for a molecular superconductor. Since then, superconductivity has been reported in fullerene doped with various other alkali metals. It has been shown that the superconducting transition temperature in alkaline-metal-doped fullerene increases with the unit-cell volume V. As caesium forms the largest alkali ion, caesium-doped fullerene is an important material in this family. Recently, superconductivity at 38 K has been reported in bulk Cs_3C_{60}, but only under applied pressure. The highest superconducting transition temperature of 33 K at ambient pressure is reported for Cs_2RbC_{60}.

The increase of transition temperature with the unit-cell volume had been believed to be evidence for the BCS mechanism of C_{60} solid superconductivity, because inter C_{60} separation can be related to an increase in the density of states on the Fermi level, $N(\varepsilon_F)$. Therefore, there have been many efforts to increase the interfullerene separation, in particular, intercalating neutral molecules into the A_3C_{60} lattice to increase the interfullerene spacing while the valence of C_{60} is kept unchanged. However, this ammoniation technique has revealed a new aspect of fullerene intercalation compounds: the Mott transition and the correlation between the orientation/orbital order of C_{60} molecules and the magnetic structure.

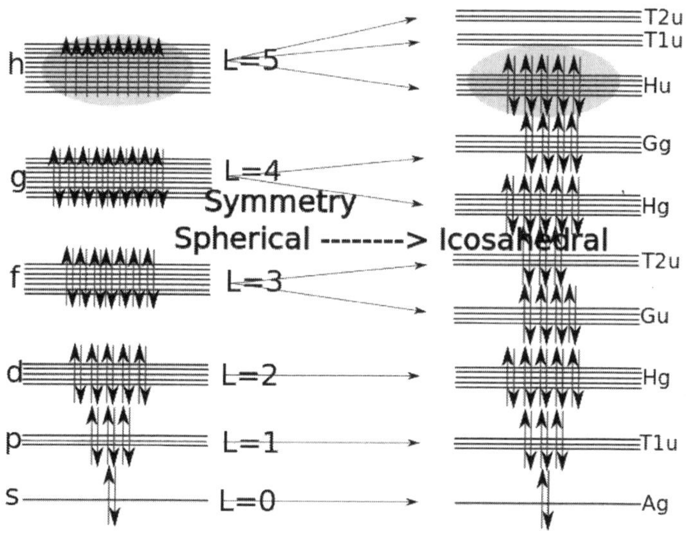

Figure: *Electronic structure of C_{60} under "ideal" spherical (left) and "real" icosahedral symmetry (right).*

The C_{60} molecules compose a solid of weakly bound molecules. The fullerites are therefore molecular solids, in which the molecular properties still survive. The discrete levels of a free C_{60} molecule are only weakly broadened in the solid, which leads to a set of essentially nonoverlapping bands with a narrow width of about 0.5 eV. For an undoped C_{60} solid, the 5-fold h_u band is the HOMO level, and the 3-fold t_{1u} band is the empty LUMO level, and this system is a band insulator. But when the C_{60} solid is doped with metal atoms, the metal atoms give electrons to the t_{1u} band or the upper 3-fold t_{1g} band. This partial electron occupation of the band may lead to metallic behaviour. However, A_4C_{60} is an insulator, although the t_{1u} band is only partially filled and it should be a metal according to band theory.

This unpredicted behaviour may be explained by the Jahn-Teller effect, where spontaneous deformations of high-symmetry molecules induce the splitting of degenerate levels to gain the electronic energy. The Jahn-Teller type electron-phonon interaction is strong enough in C_{60} solids to destroy the band picture for particular valence states.

A narrow band or strongly correlated electronic system and degenerated ground states are important points to understand in explaining superconductivity in fullerene solids. When the inter-electron repulsion U is greater than the bandwidth, an insulating localized electron ground state is produced in the simple Mott-Hubbard model.

This explains the absence of superconductivity at ambient pressure in caesium-doped C_{60} solids. Electron-correlation-driven localization of the t_{1u} electrons exceeds the critical value, leading to the Mott insulator. The application of high pressure decreases the interfullerene spacing, therefore caesium-doped C_{60} solids turn to metallic and superconducting.

A fully developed theory of C_{60} solids superconductivity is still lacking, but it has been widely accepted that strong electronic correlations and the Jahn-Teller electron-phonon coupling produce local electron-pairings that show a high transition temperature close to the insulator-metal transition.

Chemical Reactions and Properties

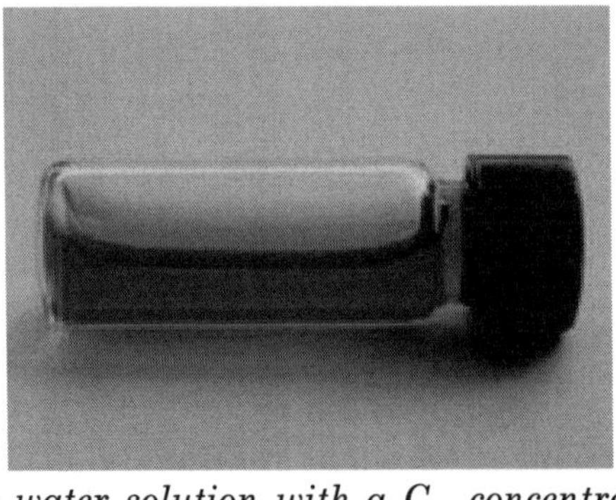

Figure: *C_{60}HyFn water solution with a C_{60} concentration of 0.22 g/L.*

Hydrated Fullerene (HyFn): Hydrated fullerene C_{60}HyFn is a stable, highly hydrophilic, supra-molecular complex consisting of C_{60} fullerene molecule enclosed into the first hydrated shell that contains 24 water molecules: $C_{60}@(H_2O)_{24}$. This hydrated shell is formed as a result of donor-acceptor interaction between lone-electron pairs of oxygen, water molecules and electron-acceptor centres on the fullerene surface.

Meanwhile, the water molecules which are oriented close to the fullerene surface are interconnected by a three-dimensional network of hydrogen bonds. The size of C_{60}HyFn is 1.6–1.8 nm. The maximal concentration of C_{60} in the form of C_{60}HyFn achieved by 2010 is 4 mg/mL.

Hydrogenation

C_{60} exhibits a small degree of aromatic character, but it still reflects localized double and single C-C bond characters. Therefore C_{60} can undergo addition with hydrogen to give polyhydrofullerenes. C_{60} also undergoes Birch reduction. For example, C_{60} reacts with lithium in liquid ammonia, followed by *tert*-butanol to give a mixture of polyhydrofullerenes such as $C_{60}H_{18}$, $C_{60}H_{32}$, $C_{60}H_{36}$, with $C_{60}H_{32}$ being the dominated product. These mixture of polyhydrofullerenes can be re-oxidized by 2,3-Dichloro-5,6-dicyano-1,4-benzoquinone to give C_{60} again.

Selective hydrogenation method exists. Reaction of C_{60} with 9,9',10,10'-dihydroanthracene under the same conditions, depending on the time of reaction, gives $C_{60}H_{32}$ and $C_{60}H_{18}$ respectively and selectively.

C_{60} can be hydrogenated, suggesting that a modified buckminsterfullerene called organometallic buckyballs (OBBs) could become a vehicle for "high density, room temperature, ambient pressure storage of hydrogen". These OBBs are created by binding atoms of a transition metal (TM) to C_{60} or $C_{48}B_{12}$ and then binding many hydrogen atoms to this TM atom, dispersing them evenly throughout the inside of the organometallic buckyball. The study found that the theoretical amount of H_2 that can be retrieved from the OBB at ambient pressure approaches 9 wt %, a mass fraction that has been designated as optimal for hydrogen fuel by the U.S. Department of Energy.

Halogenation

Addition of fluorine, chlorine, and bromine occurs for C_{60}.

For fluorine, the F atoms are small enough for a 1,2-addition. However, for Cl_2 and Br_2, the halogen atoms adds to remote C atoms due to steric factors. For example, in $C_{60}Br_8$ and $C_{60}Br_{24}$, the Br atoms are in 1,3- or 1,4-positions with respect to each other.

Under various conditions, a vast number of halogenated derivatives of C_{60} can be produced, some with extraordinary selectivity on one or two isomers over the other possible ones.

Addition of fluorine and chlorine usually results in a flattening of the C_{60} framework into a drum-shaped molecule.

Addition of Oxygen Atoms

Solutions of C_{60} can be oxygenated to the epoxide $C_{60}O$. Ozonation of C_{60} in 1,2-xylene at 257K gives an intermediate ozonide $C_{60}O_3$, which can be decomposed into 2 forms of $C_{60}O$. Decomposition of $C_{60}O_3$ at 296K gives the epoxide, but photolysis gives a product in which the O atom bridges a 5,6-edige.

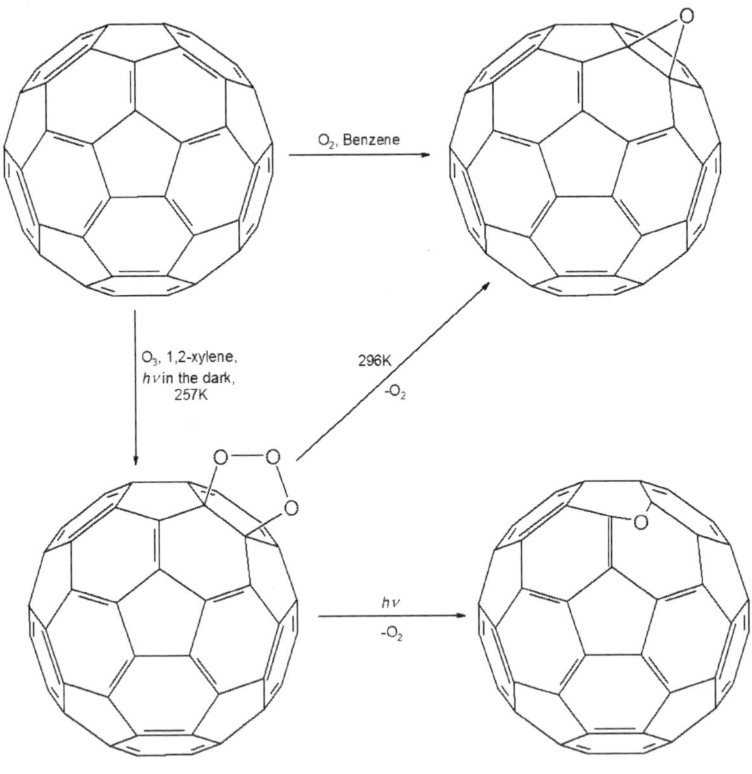

Cycloadditions

The Diels-Alder reaction is commonly employed to functionalize C_{60}. Reaction of C_{60} with appropriate substituted diene gives the corresponding adduct.

The Diels-Alder reaction between C_{60} and 3,6-diaryl-1,2,4,5-tetrazines affords C_{62}. The C_{62} has the structure in which a four-membered ring is surrounded by four six-membered rings.

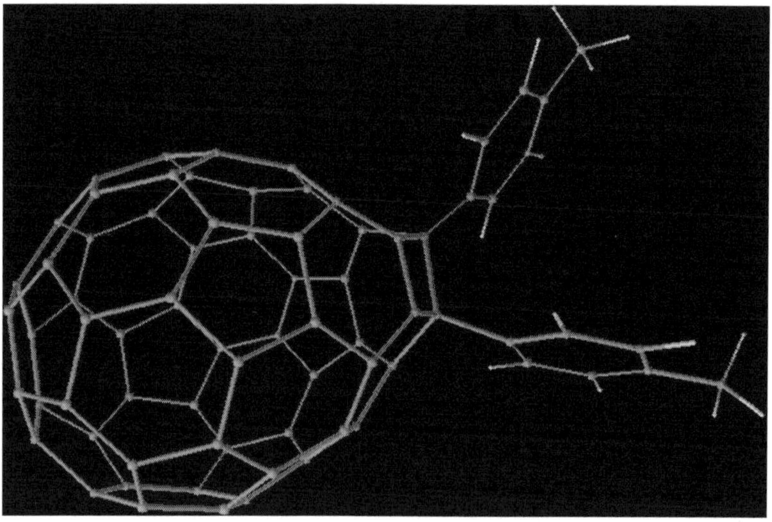

***Figure:** A C_{62} derivative $[C_{62}(C_6H_4\text{-}4\text{-}Me)_2]$ synthesized from C_{60} and 3,6-bis(4-methylphenyl)-3,6-dihydro-1,2,4,5-tetrazine*

The C_{60} molecules can also be coupled through a [2+2] cycloaddition, giving the dumbbell-shaped compound C_{120}. The coupling is achieved by high-speed vibrating milling of C_{60} with a catalytic amount of KCN. The reaction is reversible as C_{120} dissociates back to two C_{60} molecules when heated at 450 K (177 °C; 350 °F). Under high pressure and temperature, repeated [2+2] cycloaddtion between C_{60} results in a polymerized fullerene chains and networks. These polymers remain stable at ambient pressure and temperature once formed, and have remarkably interesting electronic and magnetic properties, such as being ferromagnetic above room temperature.

Free Radical Reactions

Reactions of C_{60} with free radicals readily occur. When C_{60} is mixed with a disulfide RSSR, the radical $C_{60}SR\cdot$ forms spontaneously upon irradiation of the mixture.

Stability of the radical species $C_{60}Y\cdot$ depends largely on steric factors of Y. When *tert*-butyl halide is photolyzed and allowed to react with C_{60}, a reversible inter-cage C-C bond is formed:

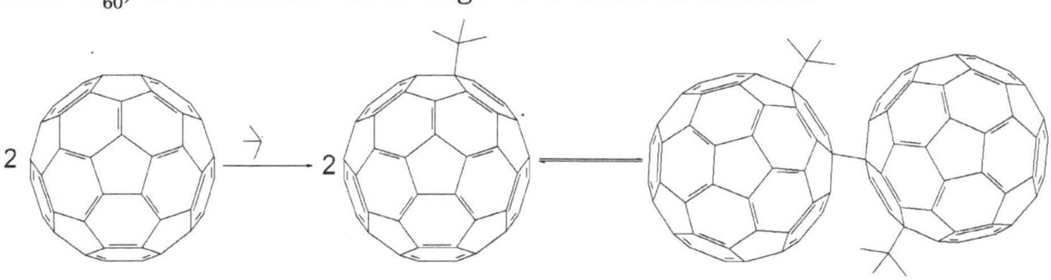

Cyclopropanation (Bingel Reaction)

Cyclopropanation (Bingel reaction) is another common method for functionalizing C_{60}. Cyclopropanation of C_{60} mostly occurs at the junction of 2 hexagons due to steric factors.

The first cyclopropanation was carried out by treating the β-bromomalonate with C_{60} in the presence of a base. Cyclopropanation also occur readily with diazomethanes. For example, diphenyldiazomethane reacts readily with C_{60} to give the compound $C_{61}Ph_2$. Phenyl-C61-butyric acid methyl ester derivative prepared through cyclopropanation has been studied for use in organic solar cells.

Redox Reactions – C_{60} Anions and Cations

The LUMO in C_{60} is triply degenerate, with the HOMO-LUMO separation relatively small. This small gap suggests that reduction of C_{60} should occur at mild potentials leading to fulleride anions, $[C_{60}]^{n-}$ (n = 1–6). The midpoint potentials of 1-electron reduction of buckminsterfullerene and its anions is given in the table below:

Reduction Potential of C_{60} at 213K

Half-Reaction	*E° (V)*
$C_{60} + e^- \rightleftharpoons C_{-60}$	–0.169
$C_{-60} + e^- \rightleftharpoons C_{2-60}$	–0.599
$C_{2-60} + e^- \rightleftharpoons C_{3-60}$	–1.129
$C_{3-60} + e^- \rightleftharpoons C_{4-60}$	–1.579
$C_{4-60} + e^- \rightleftharpoons C_{5-60}$	–2.069
$C_{5-60} + e^- \rightleftharpoons C_{6-60}$	–2.479

C_{60} forms a variety of charge-transfer complexes, for example with tetrakis(dimethylamino)ethylene:

$$C_{60} + C_2(NMe_2)_4 \rightarrow [C_2(NMe_2)_4]^+[C_{60}]^-$$

This salt exhibits ferromagnetism at 16K.

The paramagnetic fulleride ion $[C_{60}]^{2-}$ has been isolated as the $[K(crypt\text{-}222)]^+$ salt. It is synthesized by treating C_{60} with metallic potassium in the presence of 2.2.2-Cryptand. The most common fulleride ion, however, is $[C_{60}]^{3-}$.

Alkali metal salts of this trianion are superconducting. In M_3C_{60} (M = Na, K, Rb), the M^+ ions occupy the interstitial holes in a lattice composed of ccp lattice composed of nearly spherical C_{60} anions. In Cs_3C_{60}, the cages are arranged in a bcc lattice.

Critical Temperatures (T_c) of the Fulleride Salts M_3C_{60}

Salt	*T_c (K)*
Na_3C_{60}	(non-superconductive)
K_3C_{60}	18
Rb_3C_{60}	28
Cs_3C_{60}	40

C_{60} oxidizes with difficulty. Three reversible oxidation processes have been observed by using cyclic voltammetry with ultra-dry methylene chloride and a supporting electrolyte with extremely high oxidation resistance and low nucleophilicity, such as $[^nBu_4N]$ $[AsF_6]$.

Reduction Potentials of C_{60} Oxidation at Low Temperatures

Half-reaction	E° (V)
$C_{60} \rightleftharpoons C_{+60}$	+1.27
$C_{+60} \rightleftharpoons C_{2+60}$	+1.71
$C_{2+60} \rightleftharpoons C_{3+60}$	+2.14

Which the $[C_{60}]^{2+}$ ion is very unstable, and the third process can be studied only at low temperatures.

The redox potentials of C_{60} can be modified supramolecularly. A dibenzo-18-crown-6 derivative of C_{60} has been made, featuring a voltage sensor device, with the reversible binding of K^+ ion causing an anodic shift of 90mV of the first C_{60} reduction.

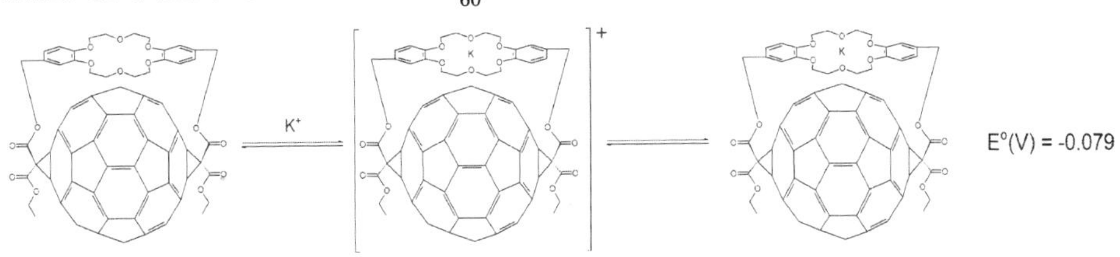

Metal Complexes

C_{60} forms complexes akin to the more common alkenes. Complexes have been reported molybdenum, tungsten, platinum, palladium, iridium, and titanium. The pentacarbonyl species are produced by photochemical reactions.

$$M(CO)_6 + C_{60} \rightarrow M(\eta^2\text{-}C_{60})(CO)_5 + CO\ (M = Mo, W)$$

In the case of platinum complex, the labile ethylene ligand is the leaving group in a thermal reaction:

$$Pt(\eta^2\text{-}C_2H_4)(PPh_3)_2 + C_{60} \rightarrow Pt(\eta^2\text{-}C_{60})(PPh_3)_2 + C_2H_4$$

Titanocene complexes have also been reported:

$$(\eta^5\text{-}Cp)_2Ti(\eta^2\text{-}(CH_3)_3SiC\equiv CSi(CH_3)_3) + C_{60} \rightarrow (\eta^5\text{-}Cp)_2Ti(\eta^2\text{-}C_{60}) + (CH_3)_3SiC\equiv CSi(CH_3)_3$$

Coordinatively unsaturated precursors, such as Vaska's complex, for adducts with C_{60}:

$$trans\text{-}Ir(CO)Cl(PPh_3)_2 + C_{60} \rightarrow Ir(CO)Cl(\eta^2\text{-}C_{60})(PPh_3)_2$$

One such iridium complex, $[Ir(\eta^2\text{-}C_{60})(CO)Cl(Ph_2CH_2C_6H_4OCH_2Ph)_2]$ has been prepared where the metal centre projects two electron-rich 'arms' that embrace the C_{60} guest.

Endohedral Fullerenes

Metal atoms or certain small molecules such as H_2 and noble gas can be encapsulated inside the C_{60} cage. These endohedral fullerenes are usually synthesized by doping in the metal atoms in an arc reactor or by laser evaporation. These methods gives low yields of endohedral fullerenes, and a better method involves the opening of the cage, packing in the atoms or molecules, and closing the opening using certain organic reactions. This method, however, is still immature and only a few species have been synthesized this way.

Endohedral fullerenes show distinct and intriguing chemical properties that can be completely different from the encapsulated atom or molecule, as well as the fullerene itself. The encapsulated atoms have been shown to perform circular motions inside the C_{60} cage, and its motion has been followed by using NMR spectroscopy.

Applications

No application of C_{60} has been commercialized. In the medical field, elements such as helium (that can be detected in minute quantities) can be used as chemical tracers in impregnated buckyballs.

Buckminsterfullerene also inhibit the HIV virus. In particular, C_{60} inhibits a key enzyme in the human immunodeficiency virus known as HIV-1 protease; this could inhibit reproduction of the HIV virus in immune cells.

The optical absorption properties of C_{60} match solar spectrum in a way that suggests that C_{60}-based films could be useful for photovoltaic applications. Conversion efficiencies up to 5.7% have been reported in C_{60}-polymer cells.

Nanofiltration

Nanofiltration is a relatively recent membrane filtration process used most often with low total dissolved solids water such as surface water and fresh groundwater, with the purpose of softening (polyvalent cation removal) and removal of disinfection by-product precursors such as natural organic matter and synthetic organic matter.

Nanofiltration is also becoming more widely used in food processing applications such as dairy, for simultaneous concentration and partial (monovalent ion) demineralisation.

Nanofiltration is a membrane filtration based method that uses nanometre sized cylindrical through-pores that pass through the membrane at a 90°. Nanofiltration membranes have pore sizes from 1-10 Angstrom, smaller than that used in microfiltration and ultrafiltration, but just larger than that in reverse osmosis.

Membranes used are predominantly created from polymer thin films. Materials that are commonly used include polyethylene terephthalate or metals such as aluminum.

Pore dimensions are controlled by pH, temperature and time during development with pore densities ranging from 1 to 106 pores per cm^2. Membranes made from polyethylene terephthalate and other similar materials, are referred to as "track-etch" membranes, named after the way the pores on the membranes are made. "Tracking" involves bombarding the polymer thin film with high energy particles. This results in making tracks that are chemically developed into the membrane, or "etched" into the membrane, which are the pores. Membranes created from metal such as alumina membranes, are made by electrochemically growing a thin layer of aluminum oxide from aluminum metal in an acidic medium.

Range of Applications

Historically, nanofiltration and other membrane technology used for molecular separation was applied entirely on aqueous systems. The original uses for nanofiltration were water treatment and in particular water softening. Nanofilters can "soften" water by retaining scale-forming, hydrated divalent ions (e.g. Ca^{2+}, Mg^{2+}) while passing smaller hydrated monovalent ions .

In recent years, the use of nanofiltration has been extended into other industries such as milk and juice production. Research and development in solvent-stable membranes has allowed the application for nanofiltration membranes to extend into new areas such as pharmaceuticals, fine chemicals, and flavour and fragrance industries. Development in organic solvent nanofiltration technology and commercialization of membranes used has extended possibilities for applications in a variety of organic solvents ranging from non-polar through polar to polar aprotic.

Industry	*Uses*
Fine chemistry and Pharmaceuticals	Non-thermal solvent recovery and management Room temperature solvent exchange
Oil and Petroleum chemistry	Removal of tar components in feed Purification of gas condensates
Bulk Chemistry	Product Polishing Continuous recovery of homogeneous catalysts
Natural Essential Oils and similar products	Fractionation of crude extracts Enrichment of natural compounds Gentle Separations
Medicine	Able to extract amino acids and lipids from blood and other cell culture.

Advantages and Disadvantages

One of the main advantages and reasons for why nanofiltration is considered a prime method of softening water is that during the process of retaining calcium and magnesium ions while passing smaller hydrated monovalent ions is performed without adding extra sodium ions, as used in ion exchangers.

Many separation processes that do not operate at room temperature (e.g. distillation), which largely increases the cost of the process when continuous heating or cooling is applied. Performing gentle molecular separation is linked with nanofiltration that is often not included with other forms of separation processes (centrifugation). These are two of the main benefits that are associated with nanofiltration.

Nanofiltration has a very favourable benefit of being able to process large volumes and continuously produce streams of have a Nanofiltration is the least used method of membrane filtration in industry as the membrane pores sizes are limited to only nanometres. Anything smaller, reverse osmosis is used and anything larger is used for ultrafiltration. Ultrafiltration can also be used in cases where nanofiltration can be used, due to it being more conventional. A main disadvantage associated with nanotechnology, as with all membrane filter technology, is the cost and maintenance of the membranes used. Nanofiltration membranes are an expensive part of the process. Repairs and replacement of membranes is dependent on total dissolved solids, flow rate and components of the feed. With nanofiltration being used across various industries, only an estimation of replacement frequency can be used. This causes nanofilters to be replaced a short time before or after their prime usage is complete.

Design and Operation

Industrial applications of membranes require hundreds to thousands of square metres of membranes and therefore an efficient way to reduce the footprint by packing them is required. Membranes first became commercially viable when low cost methods of housing in 'modules' were achieved.

Membranes are not self-supporting. They need to be stayed by a porous support that can withstand the pressures required to operate the NF membrane without hindering the performance of the membrane. To do this effectively, the module needs to provide a channel to remove the membrane permeation and provide appropriate flow condition that reduces the phenomena of concentration polarisation. A good design minimises pressure losses on both the feed side and permeate side and thus energy requirements. Leakage of the feed into the permeate stream must also be prevented. This can be done through either the use of permanent seals such as glue or replaceable seals such as O-rings.

Concentration Polarisation

Concentration polarisation describes the accumulation of the species being retained close to the surface of the membrane which reduces separation capabilities. It occurs because the particles are convected towards the membrane with the solvent and its magnitude is the balance between this convection caused by solvent flux and the particle transport away from the membrane due to the concentration gradient (predominantly caused by diffusion.) Although concentration polarisation is easily reversible, it can lead to fouling of the membrane.

Spiral Wound Module

Spiral wound modules are the most commonly used style of module and are 'standardized' design, available in a range of standard diameters (2.5", 4" and 8") to fit standard pressure vessel that can hold several modules in series connected by O-rings. The module uses flat sheets wrapped around a central tube. The membranes are glued along three edges over a permeate spacer to form 'leaves'. The permeate spacer supports the membrane and conducts the permeate to the central permeate tube.

Between each leaf, a mesh like feed spacer is inserted. The reason for the mesh like dimension of the spacer is to provide a hydrodynamic environment near the surface of the membrane that discourages concentration polarisation. Once the leaves have been wound around the central tube, the module is wrapped in a casing layer and caps placed on the end of the cylinder to prevent 'telescoping' that can occur in high flow rate and pressure conditions.

Tubular Module

Tubular modules look similar to shell and tube heat exchangers with bundles of tubes with the active surface of the membrane on the inside. Flow through the tubes is normally turbulent, ensuring low concentration polarisation but also increasing energy costs. The tubes can either be self-supporting or supported by insertion into perforated metal tubes. This module design is limited for nanofiltration by the pressure they can withstand before bursting, limiting the maximum flux possible. Due to both the high energy operating costs of turbulent flow and the limiting burst pressure, tubular modules are more suited to 'dirty' applications where feeds have particulates such as filtering raw water to gain potable water in the Fyne process. The membranes can be easily cleaned through a 'pigging' technique with foam balls are squeezed through the tubes, scouring the caked deposits.

Flux Enhancing Strategies

These strategies work to reduce the magnitude of concentration polarisation and fouling. There is a range of techniques available however the most common is feed channel spacers as described in spiral wound modules. All of the strategies work by increasing eddies and generating a high shear in the flow near the membrane surface. Some of these strategies include vibrating the membrane, rotating the membrane, having a rotor disk above the membrane, pulsing the feed flow rate and introducing gas bubbling close to the surface of the membrane.

Characterisation

Many different factors must be taken into account in the design of NF membranes, since they vary so much in material, separation mechanisms, morphology and thus application. Two important parameters should be investigated during preliminary calculations, performance and morphology parameters.

Performance Parameters

Retention of both charged and uncharged solutes and permeation measurements can be categorised into performance parameters since the performance under natural conditions of a membrane is based on the ratio of solute retained/ permeated through the membrane.

For charged solutes, the ionic distribution of salts near the membrane-solution interface plays an important role in determining the retention characteristic of a membrane. If the charge of the membrane and the composition and concentration of the solution to be filtered is known, the distribution of various salts can be found. This in turn can be combined with the known charge of the membrane and the Gibbs–Donnan effect to predict the retention characteristics for that membrane.

Uncharged solutes cannot be characterised simply by Molecular Weight Cut Off (MWCO,) although in general an increase in molecular weight or solute size leads to an increase in retention. The chemical structure, functional end-groups as well as pH of the solute, all play an important role in determining the retention characteristics and as such detailed information about the solute molecule characteristics must be known before implementing a NF design.

Morphology Parameters

The morphology of a membrane must also be known in order to implement a successful design of a NF system, and this is usually done by microscopy. Atomic force microscopy (AFM) is one method used to characterise the surface roughness of a membrane by passing a small sharp tip (<100 Γ) across the surface of a membrane and measuring the resulting Van der Waals force between the atoms in the end of the tip and the surface. This is useful as a direct correlation between surface roughness and colloidal fouling has been developed. Correlations also exist between fouling and other morphology parameters, such as hydrophobe, showing that the more hydrophobic a membrane is, the less prone to fouling it is.

Methods to determine the porosity of porous membranes have also been found via permporometry, making use of differing vapour pressures to characterise the pore size and pore size distribution within the membrane. Initially all pores in the membrane are completely filled with a liquid and as such no permeation of a gas occurs, but after reducing the relative vapour pressure some gaps will start to form within the pores as dictated by the Kelvin equation. Polymeric (non-porous) membranes cannot be subjected to this methodology as the condensable vapour should have a negligible interaction within the membrane.

Typical Figures for Industrial Applications

Keeping in mind that NF is usually part of a composite system for purification, a single unit is chosen based off the design specifications for the NF unit. For drinking water purification many commercial membranes exist, coming from different chemical families, having different structures, chemical tolerances and salt rejections and so the characterisation must be chosen based on the chemical composition and concentration of the feed stream.

NF units in drinking water purification range from extremely low salt rejection (<5% in 1001A membranes) to almost complete rejection (99% in 8040-TS80-TSA membranes.) Flow rates range from 25–60 m^3/day for each unit, so commercial filtration requires multiple NF units in parallel to process large quantities of feed water. The pressures required in these units are generally between 4.5-7.5 bar.

For seawater desalination using a NF-RO system a typical process is shown below.

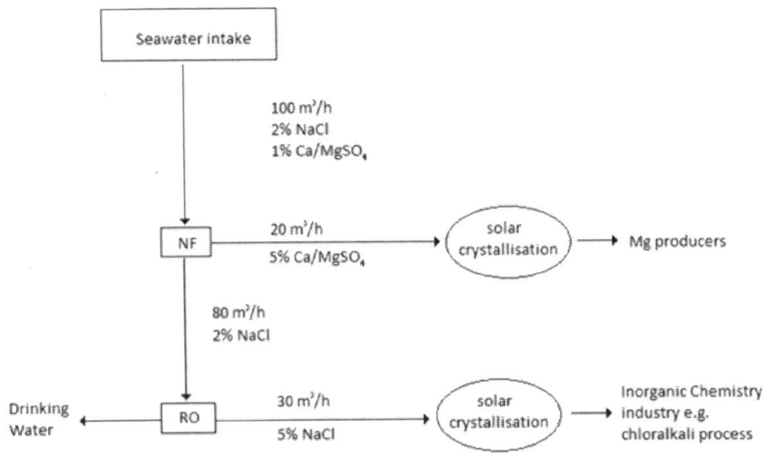

Because of the fact that NF permeate is rarely clean enough to be used as the final product for drinking water and other water purification, is it commonly used as a pre treatment step for reverse osmosis (RO) as is shown above.

Post Treatment

As with other membrane based separations such as ultrafiltration, microfiltration and reverse osmosis, post-treatment of eitherpermeate or retentate flow streams (depending on the application) – is a necessary stage in industrial NF separation prior to commercial distribution of the product. The choice and order of unit operations employed in post-treatment is dependent on water quality regulations and the design of the NF system. Typical NF water purification post-treatment stages include aeration and disinfection & stabilisation.

Aeration

A Polyvinyl chloride (PVC) or fibre-reinforced plastic (FRP) degasifier is used to remove dissolved gases such as carbon dioxide and hydrogen sulfide from the permeate stream. This is achieved by blowing air in a countercurrent direction to the water falling through packing material in the degasifier. The air effectively strips the unwanted gases from the water.

Disinfection & Stabilisation

The permeate water from a NF separation is demineralised and may be disposed to large changes in pH, thus providing a substantial risk of corrosion in piping and other equipment components. To increase the stability of the water, chemical addition of alkaline solutions such as lime and caustic soda is employed. Furthermore, disinfectants such as chlorine or chloroamine are added to the permeate, as well as phosphate or fluoride corrosion inhibitors in some cases.

New Developments

Contemporary research in the area of Nanofiltration (NF) technology is primarily concerned with improving the performance of NF membranes, minimising membrane fouling and reducing energy requirements of already existing processes. One way in which researchers are attempting to improve NF performance – more specifically increase permeate flux and lower membrane resistance – is through experimentation with different membrane materials and configurations. Thin film composite membranes (TFC), which consist of a number of extremely thin selective layersinterfacially polymerized over a microporous substrate, have had the most commercial success in industrial membrane applications due to the capability of optimizing the selectivity and permeability of each individual layer.

Recent research has shown that the addition of nanotechnology materials such as electrospunnanofibrous membrane layers (ENMs) to conventional TFC membranes results in an enhanced permeate flux. This has been attributed to inherent properties of ENMs that favour flux, namely their interconnected pore structure, high porosity and low transmembrane pressure. A recently developed membrane configuration which offers a more energy efficient alternative to the commonly used spiral wound arrangement is the hollow fibre membrane. This format has the advantage of requiring significantly less pre-treatment than spiral wound membranes, as solids introduced in the feed are displaced effectively during backwash or flushing. As a result, membrane fouling and pre-treatment energy costs are reduced.

Extensive research has also been conducted on the potential use of Titanium Dioxide (TiO_2, titania) nanoparticles for membrane fouling reduction. This method involves applying a nonporous coating of titania onto the membrane surface. Internal fouling/pore blockage of the membrane is resisted due to the nonporosity of the coating, whilst the superhydrophilic nature of titania provides resistance to surface fouling by reducing adhesion of emulsified oil on the membrane surface.

Bibliography

Altmann, J. & Gubrud, M.A.: *Discovering the Nanoscale*, IOS Press, Amsterdam, 2004.

Andrew, N.: *Foundations of Nanomechanics: from Solid-state Theory to Device Applications*, Berlin, New York, 2003.

Attebery, B.: *Nanoculture: Implications of the New Technoscience*, Intellect Books, Bristol, 2004.

Bainbridge, W.S. : *Societal Implications of Nanoscience and Nanotechnology*, Dordrecht, Kluwer, 2001.

Bast, R.C. Jr.: *Cancer Medicine*, Decker, Hamilton, 2000.

Berube, D.M.: *Discovering the Nanoscale*, IOS Press, Amsterdam, 2004.

Brosnan, J.: *The Primal Scream: A History of Science Fiction Film*, Orbit Books, London, 1991.

Carroll, J.S.: *Social Science Research Methods for Assessing Societal Implications of Nanotechnology*, Dordrecht, Kluwer, 2001.

Caviedes, C.: *El Nino in History: Storming Through the Ages*, University Press of Florida, Florida, 2001.

Credi, A.: *Molecular Devices and Machines: a Journey into the Nano World*, Wiley-VCH, Weinheim, 2003.

Crow, M. & Sarewitz, D.: *Societal Implications of Nanoscience and Nanotechnology*, Dordrecht: Kluwer, 2001.

Daniel Ratner: *Nanotechnology: a Gentle Introduction to the Next Big Idea*, Upper Saddle River, Prentice Hall, 2003.

Dobzhanshy, T.: *Genetics of the Evolutionary Process*, Columbia University Press, New York, 1970.

Drexler, K. Eric: *Engines of Creation: The Coming Era of Nanotechnology*, Anchor Books, New York, 1986.

Dyson, Tim: *India's Historical Demography: Studies in Famine, Disease, and Society*, Riverdale, Westwood, MA, 1989.

Eamonn Gearon: *The Sahara: A Cultural History*, Signal Books, UK, 2011.

Frank J. Owens: *Introduction to Nanotechnology*, Hoboken, J. Wiley, 2003.

Glimell, H.: *Dynamics of the Emerging Field of Nanoscience*, Dordrecht, Kluwer, 2001.

Glimell, H.: *Societal Implications of Nanoscience and Nanotechnology*, Kluwer, Dordrecht, 2001.

Haber, L. F.: *The Chemical Industry: 1900-1930, International Growth and Technological Change*, Oxford: Claredon Press, 1971.

Hessenbruch, A.: *Nanotechnology and the Negotiation of Novelty*, Amsterdam, IOS Press, 2004.

James Lewis: *Prospects in Nanotechnology: Toward Molecular Manufacturing,* John Wiley & Sons, New York, 1995.

Karn, Barbara: *Nanotechnology and the Environment: Applications and Implications,* American Chemical Society, Washington D.C., 2004.

Keblinski, P. : *Encyclopedia of Nanoscience and Nanotechnology*, American Scientific, Los Angeles, Calif, USA, 2004.

Liu W.K., Karpov E.G., Park H.S.: *Nano Mechanics and Materials,* Wiley, Weinheim, 2006.

Mayr, E.: *Change of Genetic Environment and Evolution*, Allen and Unwin, London, 1954.

Mody, C.C.M.: *How Probe Microscopists Became Nanotechnologists, Discovering the Nanoscale*, IOS Press, Amsterdam, 2004.

Norio Taniguchi: *Nanotechnology: Integrated Processing Systems for Ultra-precision and Ultra-fine Products,* Oxford University Press, New York, 1996.

Orth, R. N.: *Nanometer Scale Antibody Patterning for Directed cell Immobilization and Stimulation,* Cornell University, Ithaca, 2002.

Pressman, J.: *Nano Narrative: A Parable from Electronic Literature*, Intellect Books, Bristol, 2004.

Pressman, J.: *Nanoculture: Implications of the New Technoscience*, Intellect Books, Bristol, UK, 2004.

Ratner, Daniel: *Nanotechnology: a Gentle Introduction to the Next Big Idea*, Prentice Hall, Upper Saddle River, 2003.

Robison, W.L.: *Discovering the Nanoscale*, IOS Press, Amsterdam, 2004.

Sarewitz, D.: *Nanotechnology and Societal Transformation*, Dordrecht, Kluwer, 2001.

Stephen M. Goodnick: *Transport in Nanostructures,* Cambridge University Press, Cambridge, 1997.

Tenner, E.: *Nanotechnology and Unintended Consequences*, , Dordrecht, Kluwer, 2001.

Vasudev Bhatnagar: *Cell Science and Technology*, Campus Books, Delhi, 2009.

Vermaas, P.E.: *Discovering the Nanoscale*, IOS Press, Amsterdam, 2004.

Weil, V.: *Ethical Issues in Nanotechnology*, Dordrecht, Kluwer, 2001.

Index

❑❑❑